WATER JET SHIPHANDLER'S GUIDE

JOHN KINKELA

Library of Congress Control Number: 2020943456

Edited by Ian Robertson
Designed by Jack Chappell
Cover design by Jack Chappell
Type set in Pulsar/Univers/Garamond

ISBN: 978-0-7643-6140-1
Printed in India

Published by Schiffer Publishing, Ltd.
4880 Lower Valley Road
Atglen, PA 19310
Phone: (610) 593-1777; Fax: (610) 593-2002
E-mail: Info@schifferbooks.com
Web: www.schifferbooks.com

The sea is selective; slow in recognition of effort and aptitude, but fast in sinking the unfit.

—Felix Riesenberg

Always keep her in the middle of good water.

—John J. Kinkela

CONTENTS

FOREWORD

IN 2006, THE US Navy tasked me to develop a shiphandling and navigation course for the newest class of ship at the time, the Littoral Combat Ship (LCS). The course followed the Navy's new "train-to-qualify" training philosophy, which is similar to the training programs found in aviation and commercial maritime licensing. In fact the LCS program embraced a multitude of new concepts and technologies, including optimal crewing, extensive specialized simulator-based training, outsourcing of many maintenance processes, and a more robust shore support infrastructure. The physical vessels also were (and remain) cutting edge. Capable of operating at high speed and with a shallow draft, they employ lightweight construction, open computer architecture, extensive automation throughout the ship, and, significantly, water jet propulsion.

At the time there was very little modern-day experience within the Navy on handling large, water jet–propelled vessels, so we decided to go out into the civilian maritime industry to hire instructors who had professional experience with high-speed, water jet–propelled vessels and had operated in all conditions of visibility—day or night—often in very dense traffic.

Since kicking off the first class in 2006 with four students, the number of ships continues to grow and our throughput of students has increased accordingly. We have split off a second course for more-junior shiphandlers (LCS JOOD). Additionally, other Navy shiphandling courses have been based on our curriculum. In those years, I have had the good fortune of working with some excellent instructors, both civilian and military. It is a challenge to find an instructor who possesses both experience and technical skill. More challenging is finding one who is also an apt teacher. John Kinkela is both. When he first joined our team in 2014, he recognized a need for a simple, comprehensive text to teach water jet shiphandling at the novice level. For years in the commercial industry, shiphandler training for large water jet vessels was conducted "hands on" in the wheelhouse of a vessel as apprentices were observed by more-senior officers while handling a variety of situations. The Navy does not have that luxury. There are hundreds of officers coming through the training pipeline every year, and the practicalities of Navy ship employment preclude having each officer progress from zero knowledge to proficiency by simply observing what others are doing.

This book is intended to give the prospective watch stander (civilian or military) of a water jet–propelled vessel a solid basis of knowledge in the theory and practical application of water jets. With this foundation, the officer will be able to move confidently into a simulator-based or hands-on training program under the tutelage of a good instructor and quickly master the art of water jet shiphaandling.

David Kane, LCS OOD Course lead instructor
Surface Warfare Officer School, Newport, RI

PREFACE

MANY STUDENTS OF THE CLASS I teach have asked if there was a book that would outline and offer additional clarification on the concepts being taught. My own initial training made use of company literature created for in-house training. While this information was appropriate for my training, it does not translate across the water jet world. I went looking for books to recommend to my students and found little that addressed all the topics covered by the class. The books I found were either technical manuals on how the water jet itself functioned or were ship specific in nature. This book is an attempt to cover the topic so that the lessons contained are applicable to most if not all water jet–powered vessels.

The shiphandling concepts in this book were developed in real-world operations and simulators from personal experience and by master mariners who specialize in water jet shiphandling. I have been very fortunate to work with a few of these mariners, who were kind enough to let me pick their brains to write this book. Unfortunately, properly referencing each individual conversation is difficult. At the end of the book there is a list of reference books used for defining terms and concepts that are true for all ships. However, none of these books adequately address water jet shiphandling.

The intent of the book is to approach the subject material from a basic-enough point of view that it can be used by any instructional course or person who is trying to gain a better understanding of the key differences between traditional and water jet shiphandling. Additionally, there is a discussion of how maintaining the navigation watch is different at the higher speeds that most water jet vessels are capable of. Like the shiphandling topics, the concepts behind high-speed watch standing were also developed from personal experience and conversations with master mariners with high-speed experience.

It is probably best to say that while the concepts expressed in this book are as accurate as I can make them, they are my thoughts on how to best understand and address the subject. While there are other methods and ways of understanding how to control and work with water jet ships, I believe this to be the most effective way to present water jet ship handling and high-speed navigation.

ACKNOWLEDGMENTS

I WOULD FIRST LIKE to thank my wife, Maggie; without her support and understanding I would not be in a position to write this book, nor would I have had the time. Next I would like to express my gratitude to the staff at the US Navy Surface Warfare Officer school for their help, input, and corrections, including Dave Kane, Bill Lyons, and Bud Weeks. Henrick Hollesen truly helped solidify many of the concepts and explanations found in the pages that follow. Thank you to my parents, who helped start me down the path that has led to this point, and continue to be sources of inspiration and support. My sister Julianne, who was instrumental in converting my drawings to images. Last, I would like to thank all of the students who helped define the questions that needed to be answered by this book.

INTRODUCTION

THE FIRST THING TO understand is that water jet ships are handled differently than traditional propeller ships. Much of this difference centers on the way in which thrust is generated and where it is transferred to the ship. That being said, a water jet ship is a ship and, as such, will react to environmental forces such as wind, current, and shallow-water effects just as any other ship would.

The late 1800s saw the installation of water turbine auxiliary engines in European canal boats and riverboats. They used turbines similar in design to those found in hydroelectric dams, to drive water through tunnels in the sides of the boat. These turbines, while effective, were found to lack sufficient power density—the power available compared to the size of the engine—to solely power a vessel. Nevertheless, some of the elements of what would become the water jet could be found in these early designs (Author unknown, "Boat with Propelling Turbine," *Scientific American Supplement* 38, no. 980 [October 13, 1894]: 15659).

The first successful jet drive was designed by Sir William Hamilton of New Zealand in 1955. His attempts to navigate on various rivers in New Zealand resulted in damaged propulsion systems due to shallow water encountered when trying to navigate through large rapids. Deciding to mount a propulsion system that did not hang down below the hull forced Hamilton to use a high-power-density system that would not add additional draft. Initially he made use of centrifugal-style pumps, but these were bulky and inefficient. Moving away from these pumps, Hamilton made use of an improved axial-style pump. The shallow draft of his boat, combined with the high levels of thrust provided by the axial-style pump, allowed his son to become the first person to successfully navigate up the Colorado River.

From then on, jet drives were an oddity compared to the more traditional propeller and rudder combinations. Water jet–powered vessels saw limited service in various navies. These new propulsion systems were typically limited to vessels with narrow operating requirements: high speed and shallow draft. Two notable vessels from the US Navy stand out as examples: the PBR and Pegasus-class hydrofoil.

The PBR (Patrol Boat Riverine) was designed for service in the rivers of Vietnam. These waterways were known to be shallow and poorly surveyed, a combination that would present significant problems to traditional propeller and rudder vessels.

The Pegasus-class hydrofoil was designed during the Cold War by Boeing Aircraft Company to answer the need for a very fast surface vessel. They employed two different types of water jets. One was for slow-speed surface operations. The other, significantly more powerful jet was used for above-the-surface, high-speed operations.

Both these vessel types would lead to improved designs powered by water jets. Riverine patrol boats are now designed with a preference toward water jets. A significant number of water jet–powered hydrofoils continue to operate in the Far East market as passenger ferries.

While there were other vessels being propelled by water jets during the Cold War era, these were largely experimental craft and never saw regular production or use.

In the late 1970s, personal watercraft came to the market. These small one- or two-person vessels used marine versions of motorcycle engines to power small impeller-driven water jets. While limited in size and overall power, these water jets led to improvements in impeller design and function.

The late 1980s into the 1990s saw the advent of the water jet–propelled high-speed ferry. Primarily Europe and East Asia saw the greatest implementation of these new ferries, largely due to the already existing shipboard travel routes. Initially these ferries were only for passenger service. The mid-'90s saw ever-increasing vessel size, including those capable of carrying vehicles; most importantly, trucks.

The increasing number of water jet–driven ships in service has coincided with the advent of lighter, more-powerful engines. This increase has been seen mostly in the fast-ferry market, followed by various navies. In the fast-ferry sector, the introduction of jet drives has provided operators with ships capable of faster transit times and the ability to dock quickly without the use of tugs. Various navies are expressing interest due to the higher speeds this propulsion system is able to generate over conventional propeller designs, and the improved maneuvering ability and reduced draft.

The increase in numbers of small, high-powered, predominately passenger vessels throughout the world has led the UN—through the International Maritime Organization—to produce the High Speed Craft (HSC) Code, which regulates areas poorly or not covered by SOLAS regulations. High-speed craft on international voyages, many of which are water jet propelled, are required to comply with the HSC Code. The USCG and similar bodies in other countries have largely adopted these rules for vessels that do not leave territorial waters.

Contained in the code are the rules governing which vessels are subject to the code, versus the SOLAS requirements. There is a formula that takes into account the vessel's displacement and top speed to determine if a vessel should be regulated under the HSC Code of SOLAS.

The HSC Code has provisions for many things, including method, materials, and strength of construction; propulsion automation; electrical system and backups; bridge layout; crew training; allowed galley equipment; lifesaving equipment; distance to safe-haven port; and maximum safe operating envelopes for weather.

PLANING HULLS, DISPLACEMENT HULLS, AND SEMIPLANING HULLS

Ship types can fit into three categories of hull design: displacement, planing, and semiplaning. These groups can be further divided between mono- and multihulls.

Monohulls—using only one hull in the water to support the weight of the ship and her cargo—represent the largest grouping of hull type, especially among displacement-class boats. On the other hand, multihull vessels make use of two or more hulls attached by a bridging structure that remains clear of the water.

Displacement hulls have the advantage of large interior volumes at or beneath the waterline to support heavy cargoes. These hulls are often optimized for carrying large volumes or heavy weights, at the cost of obtaining higher speeds.

Planing hulls are designed so that once up to speed, the large, flat surfaces at the stern of the hull are able to support the weight of the vessel on top of the water, reducing drag. However, optimizing for speed reduces the effective amount of cargo that can be carried. Hulls of this type are small and often able to be trailered.

Semiplaning hulls combine aspects both of planing and displacement hulls to produce a hull form capable of carrying effective amounts of cargo at faster-than-typical displacement hull speeds. As with the previous categories, designing a hull this way invariably involves accepting performance trade-offs.

Another important distinction between full-displacement and semiplaning hulls is the ability to get up on plane. Planing can be defined by the hull being fully on top of the water, or "on step," with no part of the transom underwater all the way down to where it meets the bottom plating. Full-planing hulls can, with enough power, get up on plane by power alone. Semiplaning hulls get to a point where additional power being added does not lift the stern up on plane. These hulls require the use of a lift-generating device, such as a trim tab, to lift the ship up on plane.

The requirement either of planing or semiplaning hulls to be light reduces their ability to carry any significant weight. Displacement hulls are very good at carrying cargo weight, but the power requirements and fuel consumption for them to obtain high speeds are not economical for commercial operations. Certain naval ships with exceptionally high power-to-weight ratios are the primary examples of displacement hulls that are capable of high speeds.

CHAPTER ONE

WATER JET THEORY

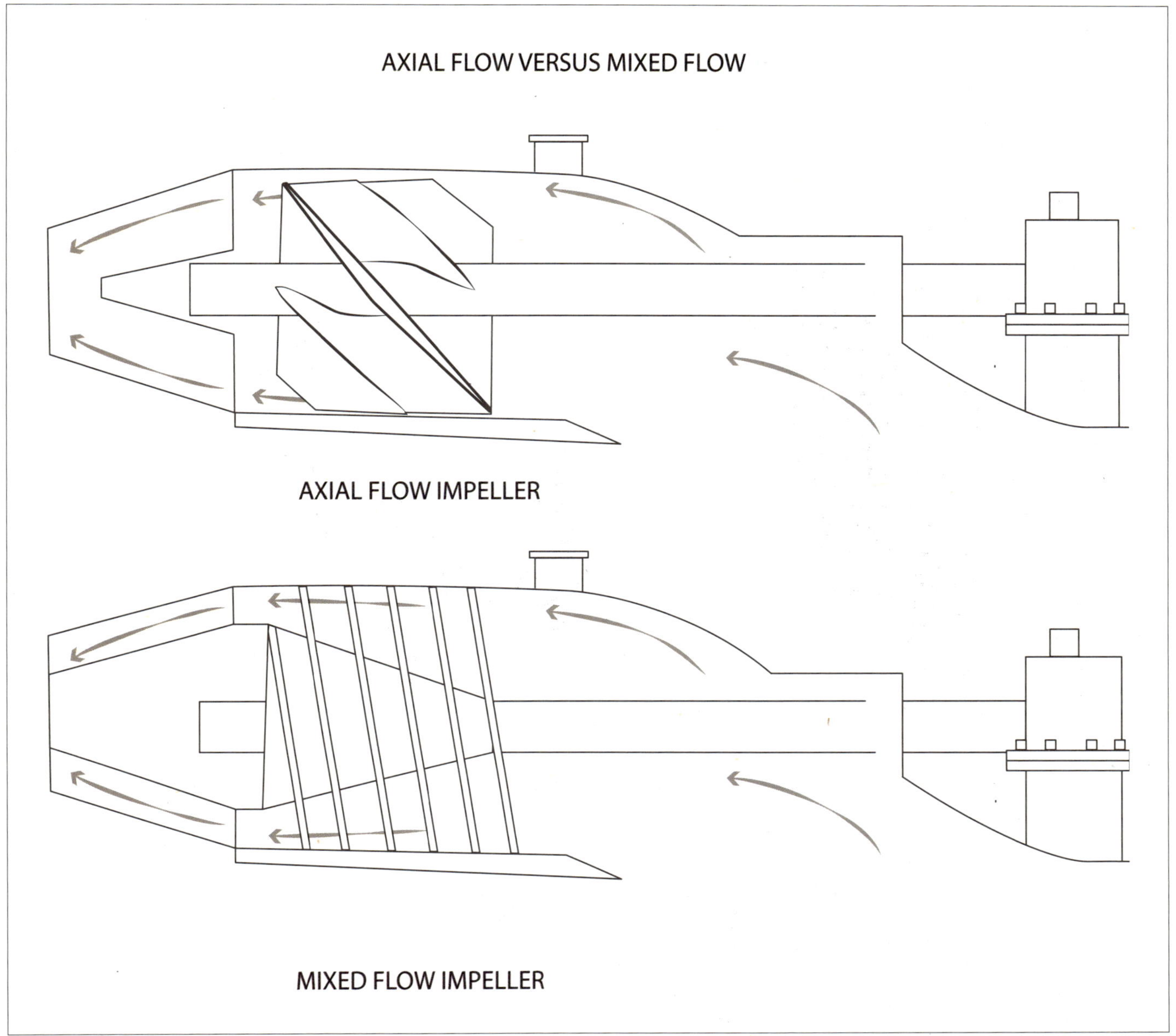

Cutaway views of axial- and mixed-flow water jets. *Courtesy of Julianne Applegate*

DESCRIPTION OF WATER JET PARTS

Water jet drives have a power plant connected by a clutch to a reduction gear that turns the shaft. For water jet propulsion, these engines are most often a high-speed diesel or gasoline engine, or a gas turbine. These engines can be fueled either with LNG or some grade of refined fuel oil. In terms of performance, gas turbines produce more power but are slower to respond to rpm changes than high-speed diesels. Medium- and slow-speed diesel engines do not have the ability to provide the required high rpm needed, or the responsiveness of the high-speed diesel to be effective powering jets for a high-speed vessel.

On vessels where the engine is a long distance from the seal—as the shaft goes from the reduction gear to the shaft seal—it is supported by a number of lineshaft bearings. These bearings are lubricated and cooled with oil, grease, or water, depending on the system. Oil pumps geared either to the engine or the reduction gears provide lube oil to the bearings, although on larger ships these oil pumps may be electrically driven. Greased bearings will use grease lubrication points and grease channels to lubricate the bearings. Water bearings use pickup tubes on the outlet side of the water jet to provide water to the bearings.

For water-lubricated bearings, there is a minimum rpm needed to provide sufficient water pressure to lubricate the bearings. Freewheeling shafts—allowing the impeller to rotate freely as water passes through the tunnel—have a critical rpm range, below which there is not enough water being picked up to lubricate the bearings. At this point, either speed needs to be increased above this threshold, or the shaft must be

Axial flow impeller from a jet ski

locked. Failure to do so may cause significant bearing damage. Below this critical rpm range there is not enough water flowing to rotate the impeller and harm the bearing.

As the size of the vessel increases, the likelihood of the engines being in the same space as the water jet machinery decreases. This auxiliary space is referred to as the jet equipment room. The jet room is where the shaft exits the hull through the shaft seal, and where the impeller is located inside the ship. As the name implies, this space will also contain the additional equipment for the water jet, such as the inspection port, hydraulics, and various electric controls.

IMPELLERS

The impeller is the heart of the water jet. The blades of an impeller are designed much differently than a propeller and work more like a pump. Typical propellers are designed to spin in open water, where there is nothing restricting the flow of water around the blades, whereas impellers spin in an enclosed space, or "housing."

An impeller is designed to spin inside a pump housing, which is considerably longer than the shroud of the Kort nozzle, requiring the impeller to draw water to the blade faces.

There are two types of impellers used in water jets: axial flow and mixed flow.

Axial flow impeller from a large ship. *Courtesy of Adam Schaulk, LT USN*

Axial-flow impellers are designed to move fluid down the axis of the shaft. This design does not reduce the area of the tunnel as significantly as a mixed-flow impeller. Unlike a propeller, the blade tip is present the full length of the blade face. The height, or chord, of the blade from hub to tip is the same while being measured from the leading edge to the trailing edge. The volume of the pump ahead of and behind the impeller is the same. These impellers produce lower pressures than the mixed flow, but the amount of water that flows through is higher. This allows the impeller to operate at higher rpm without increased risk of damaging cavitation, and higher rpm leads to higher vessel speeds.

Mixed-flow impellers are designed to move fluid out to the outside of the pump housing. A significant reduction in the area of the tunnel—caused by the hub being designed as a cone—is used to force the fluid outside. This means that the blade height reduces along the length of the blade. As the conical shape of the hub forces the water into a decreasing volume of space, it raises the pressure of the jet wash, although the flow volume will be lower than the axial-flow jet.

TUNNEL (PUMP HOUSING)

In a water jet, the pump housing is the inlet duct tunnel. A hole in the bottom of the hull forms a pipe whose outlet is the backside of the pump at the transom. The back end of the impeller is usually even with the transom. The diameter of the tunnel is not fixed for the entire length but is constricted as needed to maximize the efficiency of the impeller. This constriction is not the same as the mixed-flow jet, whose actual width of the tunnel is optimized to direct the flow of water into the impeller blades.

NOZZLE

Immediately behind the impeller is the nozzle, which serves multiple functions. The nozzle is bolted to the transom via the bolt ring, which carries the weight of the nozzle and the bucket; additionally, the after end of the shaft is supported at the center of an inner nozzle. The outer nozzle initially has the same diameter as the pump housing, while the inner nozzle will have the same diameter as the impeller hub. For mixed-flow impellers, the gap between the two nozzles will be fairly small. If axial-flow impellers have an inner nozzle, the gap between the nozzles may be large.

Stator vanes are used to connect the outer nozzle to the inner nozzle, forming the part of the nozzle called the stator bowl. These vanes are also angled so they remove the twist present in the water flow imparted by the rotating impeller. Removing the twist improves overall efficiency. After the stator bowl the nozzle begins to narrow. Continually decreasing the diameter of the nozzle from entrance to outlet increases the flow, providing yet more power to push the vessel forward. This focused force is then imparted to the hull through the transom as opposed to the thrust bearing, which is how more-conventional ships are driven forward.

The thrust nozzle is at the very back end of the outer nozzle. At this point the cross section is reduced to its smallest size, causing the force produced to be at its highest. A comparison would be a fire hose. The fire pump produces the pressure, which is then focused at the nozzle. A person holding a fire hose nozzle has to push forward against the force of the water stream coming out of the nozzle. The same force is found at the jet nozzle.

BUCKET / REVERSE DEFLECTOR

Once the rapidly flowing water leaves the nozzle, it flows into the center of the bucket, where it can be directed to control the ship. The bucket is a shroud that surrounds the end of the nozzle and can be directed left and right to vector the wash exiting the nozzle. This ability to vector the thrust is what gives water jet ships their exceptional maneuvering capability and is the secret to walking a jet ship sideways.

When one needs to move astern, the bucket's backing plate is used to redirect the jet wash forward underneath the hull. Neither the engines nor the impeller reverse direction in the normal course of operation. Paired with the backing plate is a reverse deflector that helps direct the wash forward under the hull. This deflector can take the form of ports, an additional plate in the bottom of the bucket, or an element that drops down over the nozzle exit. The reverse deflector allows the astern wash to be vectored much like the ahead wash, enabling steering while backing.

Both bucket and reverse-deflector movement is effected by hydraulic rams. These rams can be either internal or external to the transom and receive their actuating fluid from power packs in the jet equipment room. One ram is required to actuate each reversing deflector. Two additional rams are used to move the buckets side to side for steering. If adjacent buckets are mechanically linked as a set, then only two steering rams are needed to move the set. Each backing plate will still require an individual ram for the backing plate.

Inlet tunnel. *Courtesy of Henrick Hollesen*

Rear view of the nozzle. *Courtesy of Adam Schaulk, LT USN*

Front view of the nozzle. *Courtesy of Adam Schaulk, LT USN*

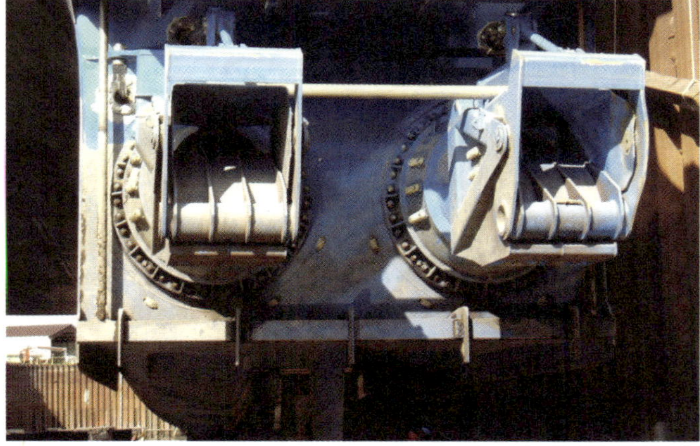

Backing plates. *Courtesy of Henrick Hollesen*

Thrust bearing

JET EQUIPMENT ROOM

The jet equipment room is the location of most of the equipment needed to operate and control the bucket and deflector. Hydraulic power packs are a combination of motor, pump, and oil sump. In some cases the motor can be replaced by a PTO (power takeoff) from the shaft to the jet pump. The control systems for directing hydraulic fluid to the rams are also in the jet room.

Many jet tunnels have an inspection port right above the impeller. This port allows for the inspection and clearing out of trapped debris in the tunnel. It is very important to carefully control when this port is opened. With the port being just at or possibly below the waterline, the potential for flooding when the hatch is open is very high. The port should be opened only after the vessel is secured to a pier, with the jet shaft secured, and with the permission of the captain. Once work has been completed, the procedure for closing and securing the port should be the same as for any other through the hull opening. An improperly secured hatch could lead to catastrophic flooding once the jet shafts are clutched in. Flooding is also very likely if the port is opened while the vessel is drifting in a seaway.

THRUST BEARING

Each shaft passes through a thrust bearing before going through the shaft seal. The thrust bearing for a propeller-driven ship takes the full force, driving the ship forward through the water, and distributes it to the hull of the ship. This requires hefty design and being well attached to the framing of the ship.

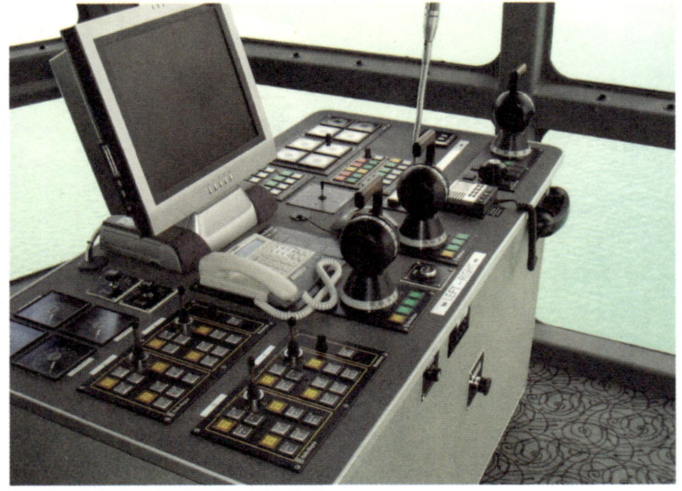

Docking console; often these consoles have not only the primary control system but also the backup. Additionally, there may be CCTV screens and other navigation and environmental indicators. *Courtesy of Henrick Hollesen*

The nozzle serves this purpose for water jet ships. The thrust bearing serves to keep the impeller properly positioned in the tunnel. Water jet shaft thrust bearings are not as heavily designed or thoroughly secured to the hull since they are not required to drive the weight of the vessel through the water.

The force driving the ship forward can be found at the nozzle and is transmitted to the transom through the jet bowl bolt ring. As the backing plate closes, the force pulling the ship aft passes from the plate to the bucket to the nozzle to the jet bolt ring. The fact that both the forward and astern forces are imparted into the ship at the same place allows multiple engines to easily cancel each other out when their backing plates are opposed.

WATER JET CONTROLS

Ships require a remote way of controlling the position of the steering and engine systems. For jet-driven ships the following must be controlled: angle of the jets, position of the reverse deflector, and engine rpm.

Many control systems make use of an electrohydraulic steering-control system to manage the actual control signals going to the steering and backing elements. These systems are very similar to the fly-by-wire systems found in modern commercial and military jets. The steering command is input into the system by the operator from a hand-steering control, by a more advanced integrated docking control, or by the autopilot. The steering-control system (SCS) then looks at vessel speed and the current power output of the engines and determines the correct amount of steering angle to send to the hydraulics at the buckets.

As the automation level increases for generating the steering command, so does the assertiveness of the SCS. While vessel is being steered by hand, the level of automation will be so low as to be almost invisible, whereas using autopilot will place the steering under total computer control.

The engines may also have their own control system running in the background (engine control system, ECS). The primary purpose of this system is to keep the engine running within safe operating ranges while protecting the engine from rapid rpm changes, excessive temperatures, and any other fault condition that could cause damage to the engine or the impeller.

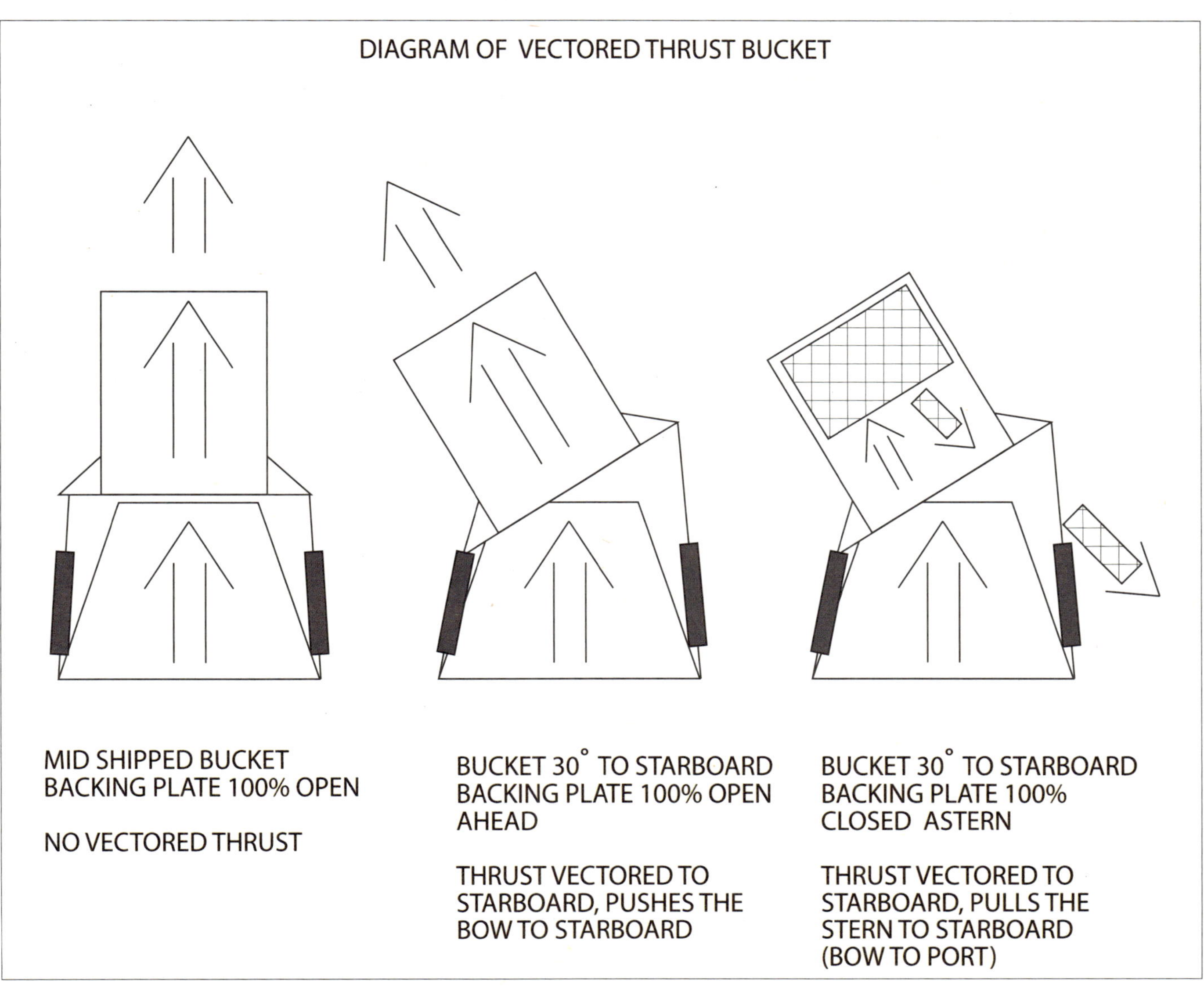

Buckets showing thrust vectoring both with ahead and astern thrust. While the center and right-side bucket are angled the same, they will steer the bow in opposite directions. The center bucket will steer the bow to starboard, and the right-side bucket will steer the bow to port. *Courtesy of Julianne Applegate*

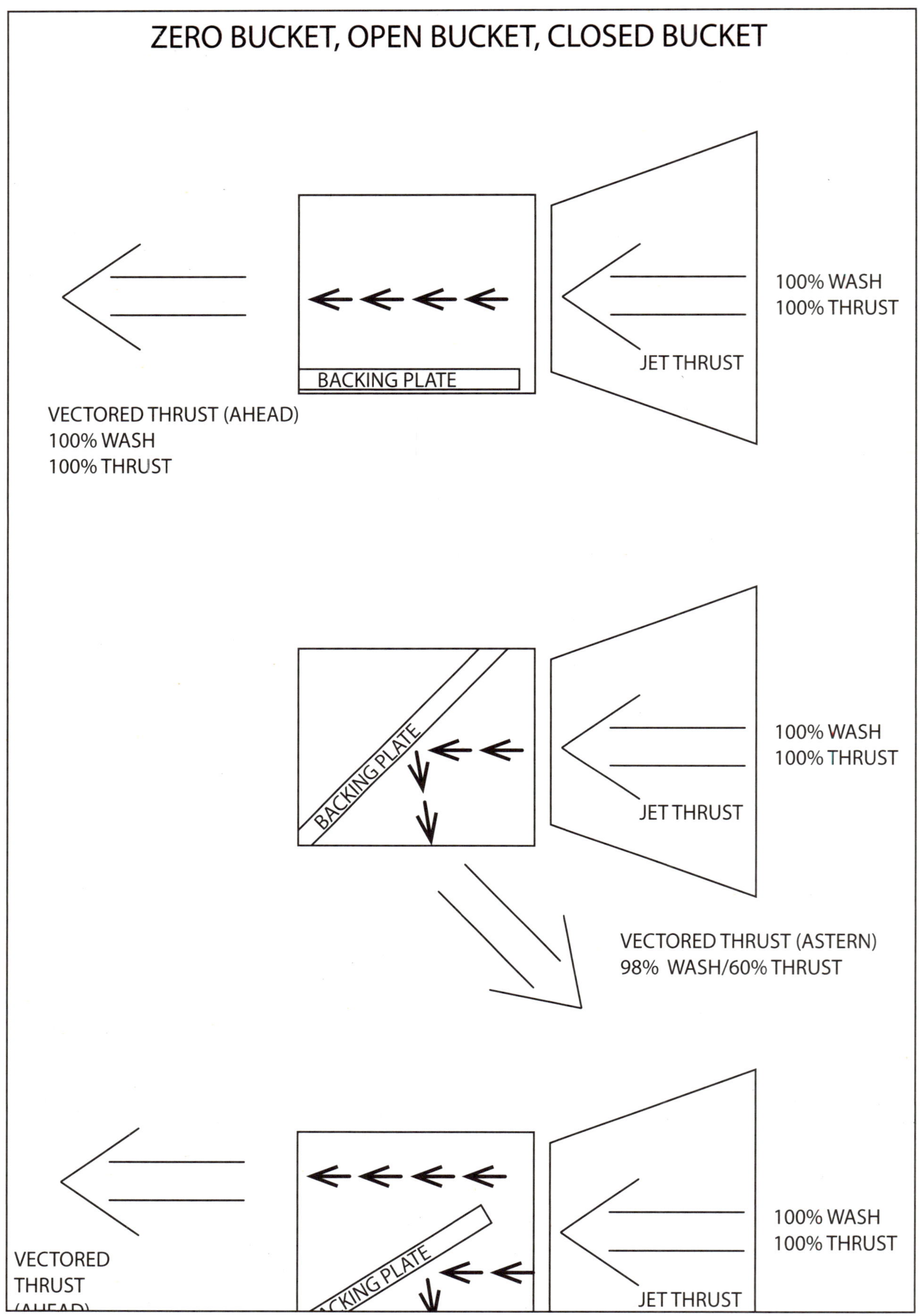

Various backing-plate positions, showing how wash from the nozzle is directed to produce ahead, astern, or zero thrust. *Courtesy of Juliann Applegate*

STEERING CONTROL: THRUST VECTORING

Rudders turn ships by being inclined to the flow of water passing around them. Increasing the angle of attack to this flow develops lift. The lift acting on the face of the rudder passes through the rudder post and up into the steering gear. Much like the thrust bearing of a propeller-driven ship, the frames supporting the steering gear and the rudder post are heavily built to transmit the turning force into the ship. This steering force can be seen as acting on the ship through the rudder post. A lever arm is created by the distance between the pivot point and the rudder post.

Water jet ships are not typically fit with rudders. Instead, steering is accomplished by vectoring the thrust coming out of the nozzle by means of the bucket. The bucket is mounted on a hinge so that it surrounds the jet nozzle. The steering rams are used to pivot the bucket port or starboard. When the bucket is offset to one side, the wash coming from the nozzle is offset in the same direction. The force of the offset wash can be seen as acting through the bucket on the jet bowl in the opposite direction of the offset while the bucket is open ahead. A bucket with ahead wash angled to port will cause the bow to go to port. A bucket with astern wash angled to port will cause the stern to go to port.

The vectored force acting on the nozzle at the transom creates a lever arm acting against the pivot point. A rudder can create this lever arm only when there is enough flow going past to produce lift. The ship must be moving forward or have a large amount of propeller wash. With vectored thrust, a water jet ship can create a lever arm while sitting still in the water, depending on pump rpm. The ability to create rate of turn (ROT) while dead in the water (DIW) is a large part of why water jet ships are so maneuverable.

FORE AND AFT MOTION: REVERSING PLATE

The reversing plate is used to quickly change the fore and aft direction of the jet wash. Much like a controlled pitch propeller (CPP), the engine of a jet drive does not need to be stopped and reversed to change the direction of thrust. However, like a CPP system, a zero-thrust position exists that is equivalent to all stop, with no outside forces acting on the ship.

When the backing plate of a water jet is set to the zero-thrust position, or "zero bucket," approximately 60 percent of the thrust is being diverted by the backing plate; the other 40 percent is allowed to exit the bucket as forward thrust. This difference actually works out to be equal due to thrust lost while being redirected by the reversing plate, and the downward angle that the reverse wash takes compared to the straight horizontal angle of the ahead wash.

There is a significant difference between the CPP zero-pitch position and the zero-bucket position—"jet creep." The zero-pitch setting for the CPP angles the propeller blades so there is no thrust or force moving the ship forward or aft. The zero-bucket position balances the jet wash exiting the nozzle so that no net thrust is produced. The two washes, ahead and astern, are producing significant amounts of thrust and force on the hull. If either of these washes interacts with a pier face or the bottom of the harbor, they may produce slightly more force on the ship, disturbing the expected balance of the zero bucket and causing a creeping motion either forward or aft. There are two primary ways to deal with jet creep: use all of the clutched-in jets to pin the ship to the pier, or clutch out any jet that is going to be left in the zero-bucket position. Any jet that is clutched in at the zero-bucket position will cause creep in either the ahead or astern direction—this is a fact of water jet propulsion. An imbalance in thrust forces caused by the backing plate's zero position being out of calibration can also cause jet creep. In this case, the solution is the same regarding clutching-out jets that are in the zero position or making use of all clutched-in jets.

Adjusting the backing plate from the zero-bucket position will produce a bias either ahead or astern. It is important to understand that while the various indicators at the conning station will show that thrust is going ahead or astern, there is still thrust going in the opposite direction of that being indicated. This bias in the thrust around each backing plate will continue until the plate is either fully opened (ahead) or fully closed (astern). The amount of bias will have an immediate effect on the speed of the vessel in the direction of that bias.

STEERING-CONTROL SYSTEMS

TILLER / WHEEL CONTROLS

Steering the vessel by hand will always be required of the bridge team. To accomplish this, vessels are fitted with either a tiller or steering wheel. These hand controls are positioned at the conning position. The most ergonomic of these controls are in the arm rests of bridge watch station chairs.

Moving the hand control will generate a steering command to the SCS. This command is processed by the SCS and sent to the steering system for all of the buckets. Since this control causes the jets to be positioned together, it is best used for normal forward steering. The position of the steering system will stay matched to the position of the control.

Some systems allow the command being sent through the SCS to reverse the signal going to one side, called steering reverse. The reversal of one steering signal will position the jets on that side, opposite of the jets on the other side. The angles of the jets will be equivalent to the position of the

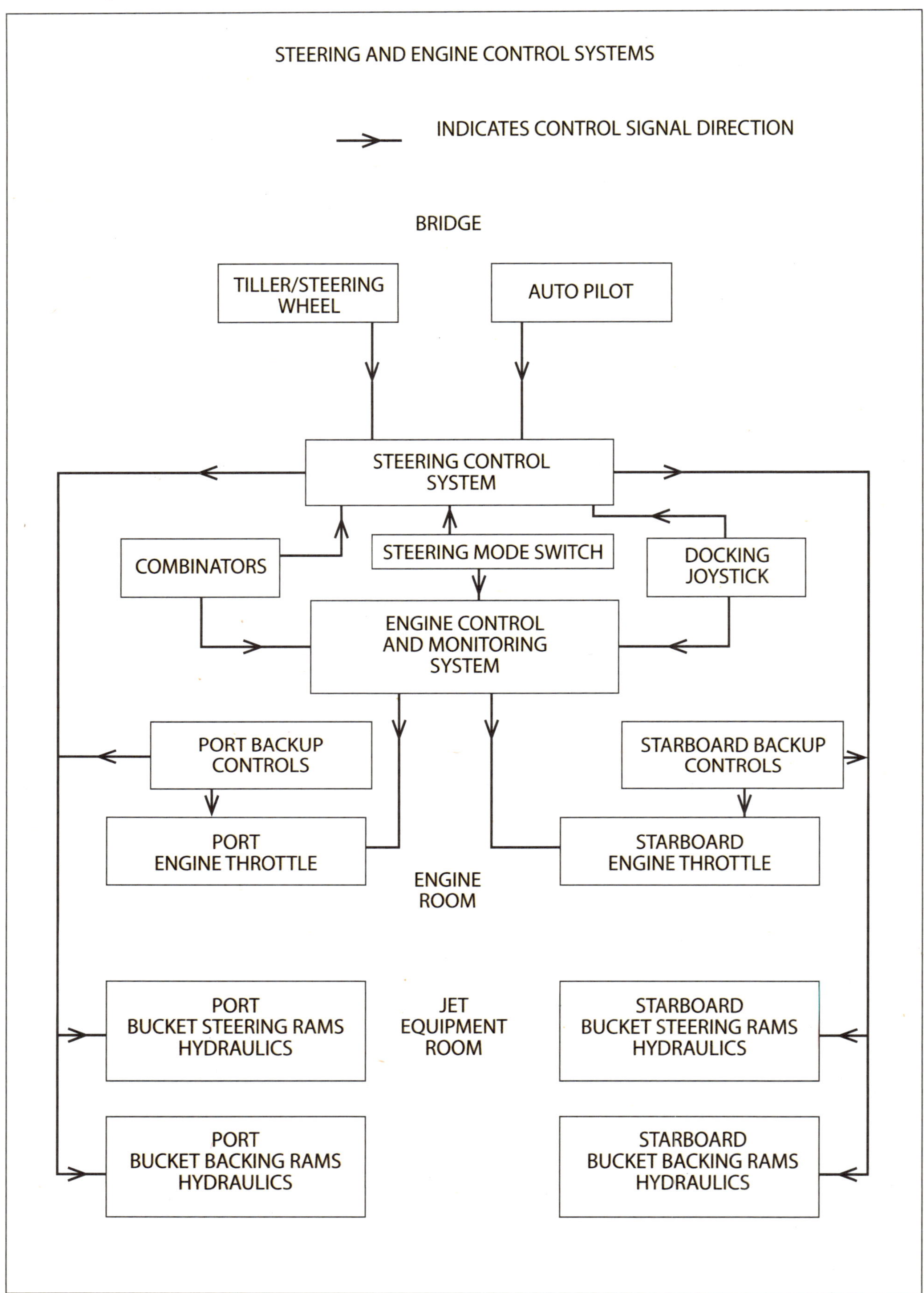

Block diagram of generic steering-control system, depicting control signals both for engines and steering. *Courtesy of Julianne Applegate*

Tiller steering control.
Courtesy of US Navy

Wheel steering control. *Courtesy of Henrik Hollesen*

control. So a command of right 10 degrees would have the effect of one side's steering jet being at right 10, while the other steering jet is at port 10 degrees. This option will have limited benefit in most cases; while I had this system on one ship, it was never used.

While operating in hand or tiller steering mode, engines are controlled—through ECS—from a throttle that is separate from the steering control.

COMBINATORS

While some vessels use a tiller or wheel arrangements for steering, others may use "combinators." As the name implies, both the steering angle of the water jet and the power and direction of jet thrust are adjusted by this one control. A vessel may be fitted with both types of controls for use in different situations. Combinators found on water jet craft are similar in function to those found in use onboard any vessel with azimuthing drives. In the case of jet drives, the rotational part of the combinator controls the angle of the steering system; the throttle portion controls reverse-deflector position and engine rpm. Combinators are provided for each steerable jet, or each set of mechanically linked jets.

The commands generated by the combinators are sent to the SCS and the ECS. During hand steering, the control systems continue to operate in the background. Some systems will allow one combinator to become the master, linking sets of jets electronically in the SCS and ECS. The remaining combinators are slaved to the master or taken completely offline. Operating this way is like hand steering, in that all the jet angles, and additionally engine rpm, are controlled with one hand.

Ships with more than two steerable jets on each side of the ship can, either mechanically or electronically, link multiple jets to one or two combinators. This allows the operator to use two combinators to oppose the direction of multiple steerable/ reversible jets, instead of trying to control more than two combinators at a time for close-quarters maneuvering. Having one set of jets thrusting astern while the other set is thrusting ahead sets up the opposed vectors that enable walking.

To induce a walk, the jets will need to be toed out. The home position will vary from ship class to ship class and is largely dependent on the distance from the centerline of the ship to the center of the bucket compared to the waterline length of the ship. The power or speed of the walk is controlled by the amount of difference between ahead and astern shaft rpm. Longer opposed vector lengths will produce a higher lateral-walking speed. In a toed-out condition, the ship will walk to the same side as the astern jet. While the ship is walking, its heading is controlled by adjusting the ahead bucket slightly from the home angle. Further discussion of this topic is in the shiphandling section.

DOCKING JOYSTICKS

Having a SCS between the controls and the jet hydraulics allows some systems to offer an integrated docking control or joystick. This control mode has the highest level of SCS and ECS involvement. A joystick control is typically at a dedicated docking station. Control inputs are sent to the SCS and ECS and are combined to control the movement of the whole ship, rather than controlling one set of jets.

For example, when the docking-control lever is moved to the 090 position, commands are sent to the SCS and ECS. This command is then interpreted and a signal is sent to the jet hydraulics, causing the jets to slew to angles determined by the SCS, while the ECS adjusts the reverse deflectors and engine rpm separately for each set of jets to achieve an equal level of power so that fore and aft motion is theoretically kept at zero while generating the forces needed to move or walk the ship sideways.

Most integrated docking systems make use of a heading-adjustment control or "moment knob." Using this control to adjust the heading will momentarily override the commands issued from the docking joystick. Commands from the joystick are immediately dropped and new signals are sent to the jets that will cause the bow to move port or starboard, effectively twisting the ship as desired by the operator. As soon as the heading control is released, the commands from the joystick once again become predominant. If the heading control is activated against the direction of the walk for a significant amount of time, it will often direct the jets to reverse their angle.

Two additional controls may be found at the docking stations: an rpm control and a sensitivity adjustment. When the rpm control is used, engine power levels are adjusted evenly as required by the command from the joystick. The sensitivity adjustment tells the control system how aggressive it needs to be while responding to the commands coming from the joystick.

Integrated docking systems offer simplicity and ease of use to the operator. There is a trade-off for this ease of use, which is limited options. The operator using a docking system has to decide whether they want to walk or change heading. Operators using combinators to control the jets more directly can do both at the same time.

Docking-control stations are most often found on the bridge wings, although they can be placed anywhere in the ship where a control station might be desired for a given operational requirement.

Switching control of the vessel from the conning station to a docking station should be done with the utmost care after considering the location and speed of the ship. The use of vessel-specific checklists at each station, open lines of communication, and permission of the captain are highly recommended. Once control of the vessel has been passed to either the docking or conning station, the operator should give all controls a slight wiggle to check that control has in fact been successfully transferred.

Combinators. *Courtesy of Henrick Hollesen*

Combinators. *Courtesy of US Navy*

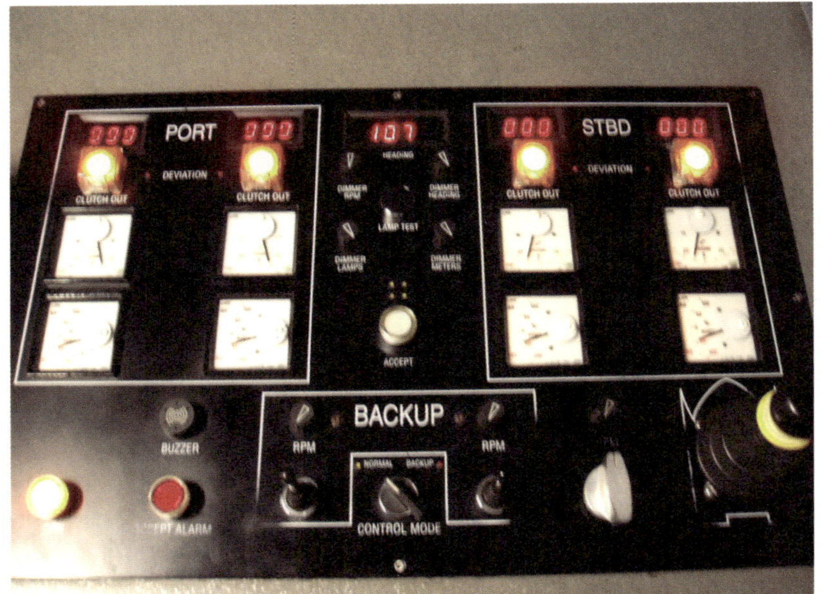

Docking joystick control. *Courtesy Henrick Hollesen*

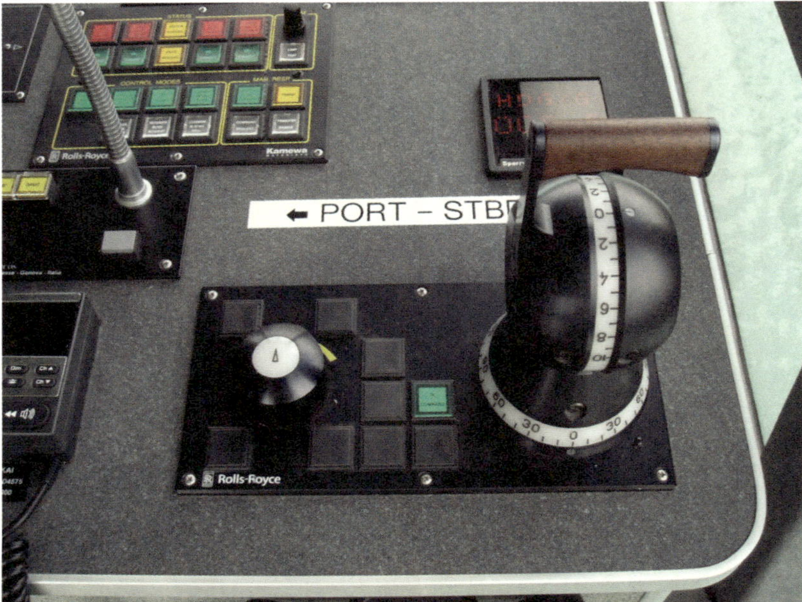

Docking joystick control. *Courtesy of Henrick Hollesen*

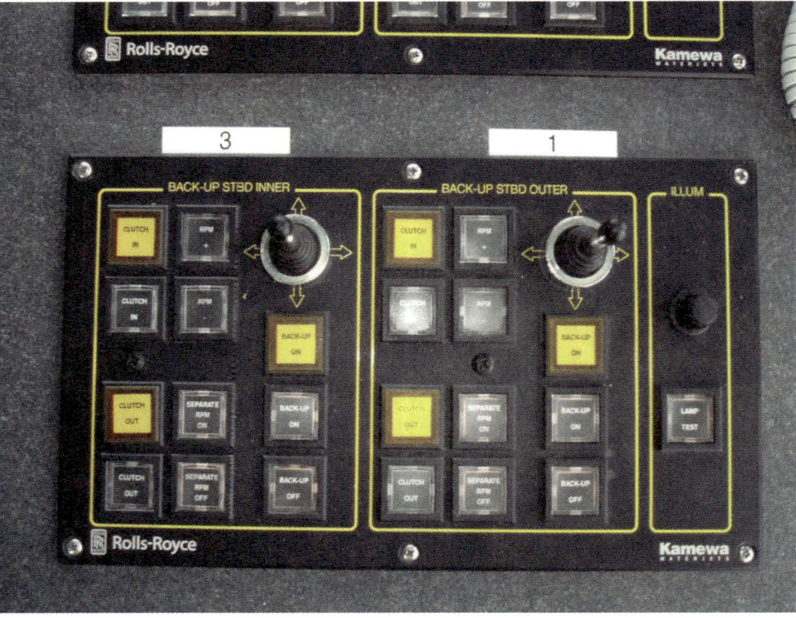

Backup control. *Courtesy of Henrick Hollesen*

BACKUP CONTROLS

The steering and engine control systems can be bypassed by use of a backup control system. The simplest and most direct way to control the jets is by backup steering. The backup controls function very similarly to the NFU (non-follow-up) control found on most ships. Just like an NFU, the position of a backup control will not match the position of the associated ram. The principal difference is that while the NFU on a conventional vessel only directly controls the rudder, the steering backup control on a water jet ship will also control the reversing deflector. The number of backups will depend on how many buckets are linked together. Typical installations have one backup for each set of port and starboard jets.

Much like the traditional NFU, moving the backup control sends a signal directly to the steering system in the jet equipment room, bypassing any other control system that may be in a fault condition.

The backup control system can be used to handle a ship alongside, but training and practice are highly recommended before attempting to do so. Any decision to operate a vessel near a dock or other obstruction on backups alone should be made with great care.

Like any other steering system, there will be a way to manually direct hydraulic fluid to the desired rams in the jet equipment room. These systems are often referred to as emergency steering systems. They are designed mainly to be used to center the jets and open the buckets in an emergency or dry dock, and not to steer the ship or in attempting to dock. The fine control needed for complex maneuvers is often not obtainable with these systems.

There are also backup controls for the engines. These are usually a simple control of the engine's rpm. Like steering backups, the rpm is directly controlled by the backup, bypassing any other control system that might be in a fault condition. Use of the engine backup should be based only on advice from the engineers responsible for engine maintenance. Without the engine control system, it is possible for the operator to unintentionally damage the engine, even while conducting normal operations.

CAVITATION: DESCRIPTION AND MANAGEMENT

Cavitation is defined as small volumes of water rapidly losing pressure as they come in contact with the impeller blade face. When this occurs the volume of water flashes to steam and expands. As it expands it rapidly cools, and the steam bubble can no longer push against the surrounding water. As quickly as the pocket of steam forms, it collapses, allowing the water to rapidly crash into the surface of the blade. The water crashing into the surface can be imagined as a wave slamming into a seawall; anything existing between the wave and the wall would be destroyed. In this way the blade surfaces are worn down at the microscopic level, known as "cavitation corrosion." With enough time and severity, it is possible to completely destroy a structure exposed to cavitation.

Water pressure is not constant across the entire surface of a traditional propeller. This is why propellers will cavitate at their tips while the rest of the blade is still developing thrust. Because impellers function inside a pump housing (the tunnel), this forces water to the blade tips and results in more-uniform pressure across the blade face.

This more-uniform pressure causes cavitation to develop across the entire face of an impeller blade, with similar causes but more-damaging effects than those on a propeller. The tight tolerances inside the tunnel are such that the effects are greatly increased compared to those seen on propellers. An impeller can show signs of cavitation across the entire front face of the blade, largely concentrating along the blade tip and root. In extreme cases, signs of cavitation will also be seen on the tunnel walls.

As cavitation eats away at the various parts of the impeller, a drop-off in performance will be observed. Reductions in performance due to cavitation will lead to more cavitation unless operating parameters are adjusted to account for the increased likelihood of cavitation. Ignoring the issue will lead to a vicious cycle where cavitation creates more damage and more cavitation.

Unfortunately there are no effective cavitation sensors to help monitor the problem; it is up to the operator to avoid cavitation. Navigation and engineering officers need to operate in a way that minimizes cavitation. This is just one of the many reasons that engineering-operating consoles and watch standers are on the bridge. By making use of a cavitation table calculated for the specific water jet, speed through the water, and ordered engine speed, the watch team can determine if cavitation is happening or not. If the watch team believes that cavitation is occurring, the only solution is to reduce the rpm of the impeller.

It is critically important to understand that not only are cavitating water jets suffering damage, they are also losing thrust and therefore cause the ship to lose steering.

Cavitation tables have a spot where the cavitation line crosses the 0-knot axis. This represents the maximum rpm the pump is capable of when not moving through the water without cavitating. This point on the table is critical to maneuvering the ship alongside a pier without damaging the impeller and is referred to as zero speed through the water (ZSTW point).

Because impellers are nondisplacement pumps, extreme cavitation will cause thrust breakdown, a condition where all movement of fluid ceases. Correcting thrust breakdown in a regular, nondisplacement pump requires repriming. This can be achieved by increasing the pressure of the water entering the pump or reducing pump rpm. It is not possible to send more water into the tunnel. The only way to stop the cavitation is to reduce impeller rpm back into the continuous operation area of the cavitation table. Once the pump has been reprimed, rpm can be increased while taking care to stay out of the no-operation zone of the cavitation table.

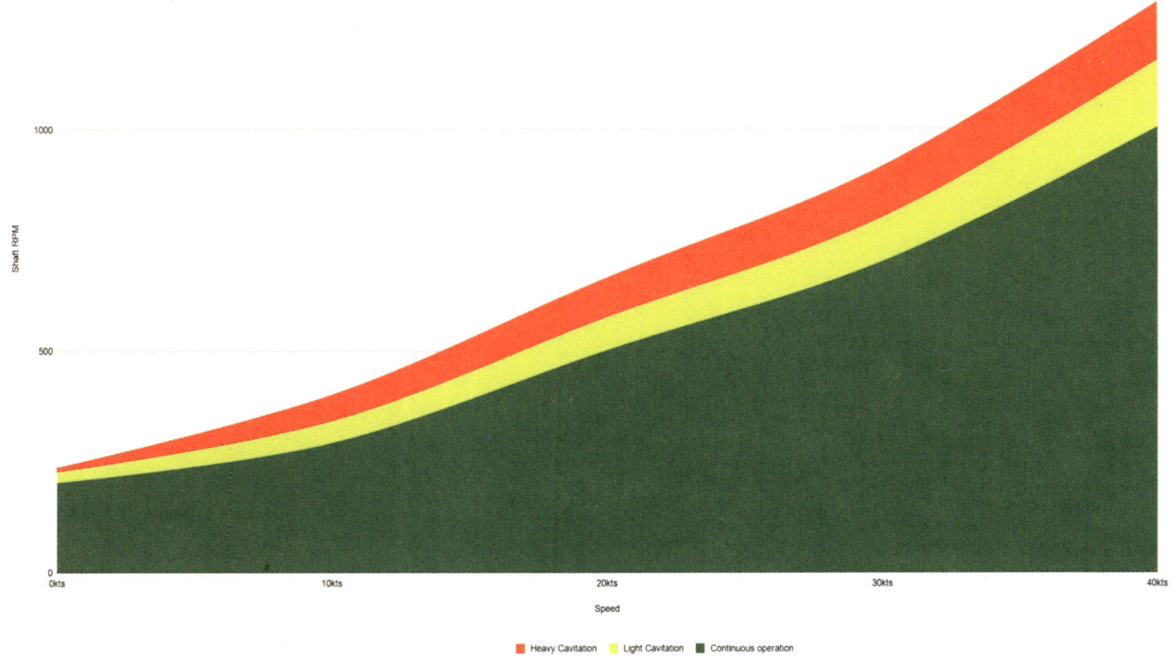

Generic cavitation table showing continuous-operation, incidental-cavitation, and no-operation zones.

Cavitation will occur any time the ratio of impeller speed to inlet flow is exceeded, referred to as pump overspeed. This ratio imbalance can be caused by running the impeller at too high an rpm, or by reducing the flow being fed to the pump.

Cavitation most often happens while accelerating, where the supply flow to the impeller is lowest and the potential to overspeed the impeller is highest. As long as vessel speed continues to increase at an appreciable rate, it can be assumed that there is no cavitation. A pump suffering thrust breakdown would not be able to continue accelerating the ship. The best way to prevent cavitation while accelerating is to advance the throttle slowly so as not to outpace the flow entering the tunnel. This will prevent inadvertently overspeeding the impeller.

Cavitation caused by reduction of flow instead of increased rpm is harder to anticipate. There are two primary situations that can reduce flow: a blockage in the inlet tunnel or insufficient water depth. These situations will be discussed at length in the following sections.

There are some common symptoms of cavitation that can help identify the situation, other than using cavitation tables. Most commonly the ship will begin to bounce in the water. This bouncing will increase as the cavitation becomes more severe. Another cavitation symptom is reduced maneuvering ability caused by the loss of wash. The steering will not be as sharp, or ordered speed will not match actual speed. If there is a view of the jet wash, it will not appear as powerful.

Cavitation is not similar to the critical speed found in many engines, and the two should not be confused.

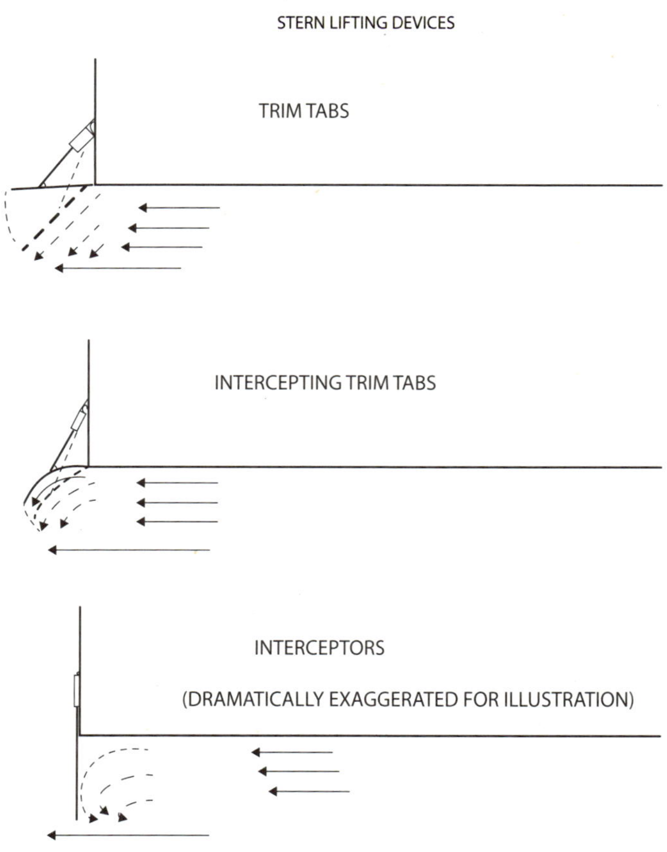

Trim tabs and an interceptor diagram. *Courtesy of Julianne Applegate*

BACKFLUSHING

When cavitation is accompanied by extreme vibration, it is likely that water flow to the impeller is being blocked by an object caught in the impeller. While impellers are designed to chew through many of the things that may get ingested into the tunnel, it is still possible for something to get stuck in the impeller that will restrict the flow of water going through the pump and imbalance the shaft.

It is hard to specifically define what objects can get stuck in the tunnel or lodged between the impeller blades. Typically these objects will be about the same density as water. Too light and they would stay on the surface; too heavy and they would fall out of the water column. Larger objects can be stuck, closing off the inlet to the tunnel. If these objects make it far enough in and come in contact with the impeller shaft, they will eventually be worn away and go through the jet. Some of these smaller bits may get wedged between the blades, preventing them from being chewed up and passing fully through the jet.

The best indication of a blockage in the tunnel or impeller is reduced thrust exiting the bucket, and this can be best observed by looking at the wash from the stern, with a good view of the jets. Blocked jet wash will look significantly different from a healthy one.

Single-jet vessels will likely have the ability to reverse the direction of the impeller shaft. This will serve to reverse the flow through the tunnel. Using this reversed flow, it may be possible to flush an obstruction back out the entrance of the inlet tunnel; this is referred to as "backflushing."

Multijet vessels are not typically equipped with reversing shafts. Backflushing the blocked jet is accomplished by backing the ship with the other, unaffected jets.

Every vessel type will have a specific backflushing procedure that takes into account the differences in hull form, bucket design, reduction gear, and clutch arrangements for that type. Vessels without reversing shafts will have a few common steps.

The affected impeller must be stopped. This may be accomplished by simply declutching the shaft. Some ships will require stopping the shaft's prime mover. Whatever the specifics surrounding the shaft, it is impossible to backflush a jet whose impeller is still running.

The bucket for the affected impeller must be fully opened, full bucket ahead. Opening the backing plate will allow water to enter the after end of the nozzle without obstruction. Many control systems will not allow the buckets to move away from zero without being clutched in. If this is the case, there should be an override control on the bridge console that allows for this; otherwise, local control will need to be established in the jet room. If the reversing bucket is allowed to remain partially closed, the effectiveness of the backflush will be reduced.

Once the impeller has been stopped and its bucket opened, the remaining jets will be placed into full bucket astern. This will back the ship, increasing the pressure of the water on the stern. The water will begin to flow into the nozzle and impeller, then out the entrance of the tunnel. The faster the ship is backed, the greater the force of the backflush, helping to push more-stubborn obstructions out. While the ship is backing, the water is flowing through the impeller, causing it to spin in reverse. For some ships, freewheeling the shaft will not present any problems and is in fact desirable. A freewheeling shaft presents the least resistance to the backflush, and the shaft rpm counter may still indicate. There are some drivetrains that cannot ever be allowed to spin in reverse. These shafts must be locked or barred from spinning before the ship is backed. Requirements for locking the shaft should be indicated in vessel-specific backflushing instructions or found in the operating manuals.

The rpm on the blocked impeller should be monitored during the backflush. An obstruction will cause the amount of water going through the tunnel to be reduced, thereby reducing the rpm. Once this blockage is cleared, the amount of water passing through the impeller will increase, causing the rpm to also increase. This is a good indication that the backflush has worked. This is not the time to stop backing the ship; continue to back the ship for at least two or three additional ship lengths. This will prevent the other jets from pulling the obstruction back in.

All efforts must be made to avoid driving right back over the obstruction. Once the affected jet is placed back into normal operation, make sure to significantly change the vessel's heading before proceeding. All impellers should be monitored for extra vibration or cavitation after backflushing.

Depending on the location of each jet inlet, it is possible for an obstruction flushed out of one jet to then enter the inlet of another jet. This would require another backflushing procedure for the newly affected jet. If backflushing efforts are not successful, the shaft should be secured until it is safe to open the inspection plate. It may be possible to pull small obstructions through the opening. Larger obstructions are more efficiently cleared by a diver, if one is available.

TRIM TABS / INTERCEPTORS AND RIDE CONTROL

As mentioned, semiplaning hulls require a lifting device on the stern to help get the hull up on plane. There are two different systems for doing this: trim tabs and interceptors.

Trim tabs are commonly found on many recreational semiplaning hulls. The tabs are positioned on the outer corners of the transom, just above the bottom plating, and are actuated by hydraulic rams mounted on the transom. The tabs push down on the water passing under the hull to lift the stern up. These tabs can be flat or curved and are the simplest way to produce the lifting force required for planing. They also produce high levels of drag. With the proper control systems, trim tabs can be used as part of the ride control while they are lifting the stern.

Interceptors, while not common, are a low-drag way to provide lift at the stern. Vertical plates are mounted across the width of the transom, just above the bottom plating. These plates are vertically actuated by hydraulic rams to slide down into the water passing under the hull. The plates produce an area of high pressure similar to that found under an airplane wing while landing. This area of high pressure lifts the stern up. Drag is significantly reduced by using interceptors, and they are equally as effective as trim tabs when used as part of ride control. The result is high levels of lift on the stern, with minimal drag. The more extended the interceptor is, the more lift it will develop.

Ride control is an active system that makes use of various control surfaces to dampen the motion of the hull in a seaway. Ride control systems are not capable of removing all of these motions and at higher sea states may not be effective at all. A properly adjusted ride control system can also help the vessel find an extra knot or two, depending on conditions. Fuel consumption is also affected by ride control settings. As ride control is set to be more aggressive, drag and, therefore, fuel consumption increase. By keeping the ride control set to just the needed level, fuel usage can be kept to a relative minimum.

For ride control to be effective there must be water flow passing over the control surfaces. Much like a rudder, without sufficient water passing by there is no effect on the ship when the control surfaces are moved, since no lift is developed. Ride control surfaces are designed to be effective at normal cruising speed. To be effective at speeds below normal cruise speeds, the various control surfaces would be oversized, causing more drag at cruising speed. Ride control is not typically effective below planing speeds.

Ride control surfaces come in many different configurations. Some vessels may have centrally mounted foil/wing forward, and others may have separate foils on each side of the bow. Vessels may also be fitted with trim tabs / interceptors on the stern or separate foils/wings on each side of the stern. Some vessels may not have any ride control surfaces forward. Most ride control systems make use of hydraulic systems to actuate the control surfaces. In the end, ride control systems are designed specifically for the vessels they are installed on, and can vary significantly from one hull to the next. Some features may be common, but exact settings will

Interceptor. *Courtesy of Henrick Hollesen*

Trim tab

be specific to the vessel and her loading condition. The manual for the specific ride control system installed on each ship should be reviewed for proper operation.

Much like the steering and engine control systems, there is a computer controlling the ride control system. An operating panel or page in the engine computer will allow operators to set system parameters such as pitch and roll angles, and how hard the system will work against the seas. The farther these settings are set from the vessel's natural state while alongside the dock will increase drag, reducing speed and fuel efficiency. If the vessel has a natural 2-degree trim, setting the system to zero trim will cause the system to work harder, reducing top speed and increasing fuel consumption. Asking the system to further reduce the effects of the ocean and smooth out the ride will also negatively affect speed and fuel consumption.

In case of an automation or sensor failure, there should be a backup method for controlling the ride control system. Unlike the jets, the vessel can be operated with the ride control system in backup or static mode, although it will not be as effective as active mode. The best way to operate in backup mode is to set the individual ride control elements to the neutral position. The elements can then be moved individually in small increments after observing vessel motion, to improve the ride. Just leaving the various elements in their zero or neutral positions is probably easiest.

Ride control fins. *Courtesy of Adam Schaulk, LT USN*

CHAPTER TWO

WATER JET SHIPHANDLING

Whether a displacement, semiplaning, or even full-planning hull is being discussed, the key item to understand is that a ship is a ship is a ship. They will all be affected by wind and current. They will all have a pivot point that moves forward or aft as speed and direction through the water changes. They will all develop bottom effects and squat. Even the wide array of propulsion choices does not change this fundamental concept. The important differences are how and where propulsion forces are directed and how ships react to them.

RATE OF TURN VERSUS VESSEL SPEED

Ships with rudders can be expected to have similar turning diameters at a wide range of speeds, assuming that engine speed is not being increased or decreased while turning. The "Standard Rudder" command is often used when a consistent radius turn is desired, no matter the vessel's speed.

Generally a water jet ship's turning diameter will increase as speed increases. The initial diameter at slow speed is significantly tighter than found on prop and rudder ships. However, at top speeds, turning diameter may be larger than similarly sized traditional vessels. As a water jet ship increases speed through the water, the available thrust turning the ship increases and maximum rate of turn will also increase, along with the final diameter of the turn. In fact, the diameter of a turn will be larger as speed is increased for the same rate of turn. The amount of increase is largely dependent on hull configuration and jet placement. Multihulls will track through the water instead of sliding across the water in a turn as monohulls do. Jets positioned close to the sides of the ship will produce more turning force for the same bucket angle.

Constructing a maneuvering table from builders' trials can be very difficult, given that turning diameters will change with differing jet combinations. The jets being used to steer, which jets are engaged, and pumping water will significantly change maneuvering characteristics. Each watch officer of a water jet ship needs to develop an understanding of how their vessel turns, with a certain amount to caution. If there is any question as to how well a jet-driven ship will turn at a given speed, it is far better to start with larger bucket angles, easing as the turn tightens more than needed. Even though this may result in high rates of turn, it is always easier to check a turn started too aggressively than to turn tighter halfway through a turn.

A note of caution for vessels operating in passenger service is required. While a higher rate of turn at high speeds is not a particular concern for safe steering and maintaining control of the ship, the lateral speed and g-forces that can be felt on the ship may cause passenger discomfort and possibly injury. It is not advisable to experiment with high rates of turn at high speed with passengers onboard.

Water jets can be pumped in a similar fashion to a rudder. With the power-to-weight ratio available in water jet ships, pumping will cause speed to increase much more quickly than on a traditional ship. Typically when pumping the jets, rpm is reduced as soon as any increase in speed is seen. This will serve to keep the speed off the vessel while holding the diameter of the turn as tight as possible.

Conversely, reduction of impeller rpm will cause an effective loss of steering. Water jet ships primarily steer by means of thrust vectoring. If the strength of the vectored thrust is reduced to a point where it no longer is capable of moving the stern, steering will not be possible. This loss of steering is similar to falling below bare steerageway. For water jets, this can happen at any speed. Any reduction in rpm will result in a degradation or loss of steering. It helps to think of the thrust and vessel speed as a ratio.

The force of the thrust is compared to the force of the water flowing past the hull as a ratio. When thrust force is the larger side of the ratio, steering will be possible. As this vector is reduced, the ratio approaches a point of balance. At this point, steering effect can be described as reduced. Steering is still possible, but higher angles and lower turn rates will be expected. A negative ratio, where thrust force is the smaller side of the ratio, results in being unable to steer the vessel in any way. Even a balanced ratio can result in the complete loss of steering.

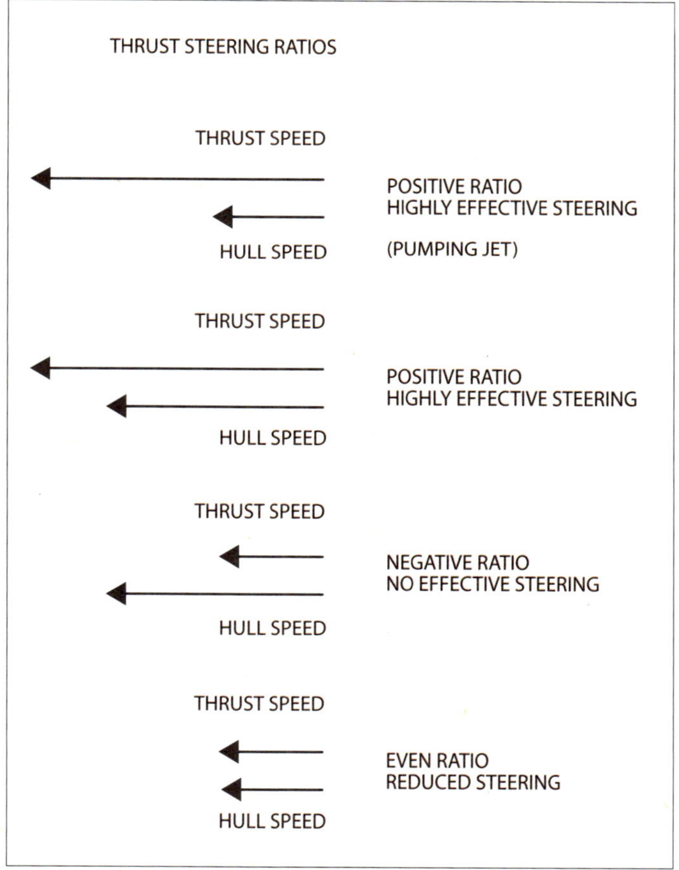

Various thrust-to-speed ratios and their steering effectiveness. *Courtesy of Julianne Applegate*

There are two ways of returning to a positive ratio and regaining steering control: wait for the vessel to slow so that thrust is the larger side of the ratio, or add thrust to make the ratio positive again.

OTHER THINGS THAT CAN AFFECT TURNING DIAMETER

Boost jets, which provide only forward nonvectored thrust, will increase the diameter of any turn when they are online; however, vessel speed will higher.

The increase in diameter that corresponds to speed in water jet ships is due to the tendency of the vectored thrust to push the stern sideways through the water while still pushing the vessel forward.

Speed will be reduced the longer the jets are angled away from centerline. This drop in speed will become sharper as the impellers come closer and closer to a cavitation condition. Once the jets start cavitating, speed will sharply fall off, causing steering effectiveness to reduce, along with a loss of propulsion.

Infrequently, water jet ships are equipped with rudders. Depending on hull layout, the rudders may increase draft when compared to a similar vessel without them. If these rudders are sized so that they are useful at all speeds, it will lead to high levels of drag when operating at planing speeds.

Rudders that are designed to be used only when up on plane are too small to be effective at slower speeds. These high-speed rudders should be positioned amidships when not planing to reduce drag as much as possible.

A ship with rudders may see a benefit in fuel consumption by setting the buckets amidships and steering with the rudders only. However, it should be noted that with steerable jets, there is no need to even have rudders installed, thereby reducing vessel weight and drag. This alone could result in the same fuel savings being sought by installing the rudders. Depending on the rudder and jet steering system, it may be possible to use both rudders and jets to steer. This should allow tighter turns than possible with jets or rudders alone.

OPERATING IN AN OPEN SEAWAY

The HSC Code will need to be considered prior to crossing an open sea that is not on the ship's regular route. The distance between safe-haven ports must be considered, as well as the expected weather. If the maximum distance to safe port or expected weather exceeds regular limitations placed on the vessel, the trip may not be possible unless a waiver is granted by the regulatory body of the flag state.

Exceeding the maximum safe-operating envelope is not to be taken lightly. High-speed vessels tend to be shallow draft and low displacement. They will not be able to handle heavy seas in the same manner as large-displacement-hull ships. In regular service, increased weather will typically result in a delayed or canceled sailing. Worsening weather when crossing the ocean will leave the crew with limited options, often requiring diverting to a safe haven and waiting for the weather to blow over. If the safe-haven port is not an option, it is best to slow and put the seas just off the bow like any other ship, even though most other ships will still be able to proceed in the same conditions.

Steering a high-speed vessel in an open seaway is very similar to any other ship. The key differences are in the ship's ability to continue to track straight and its reaction to increasing seas.

Like any other ship operating out at sea, autopilot is highly recommended for high-speed craft. High-speed hulls have little to no underwater structures, such as rudders or keels, to aid in direction keeping. The only way they maintain heading is by using the buckets to direct the wash from the jets. Not having those various structures allows the vessel to achieve very high rates of turns in excess of 100 degrees per minute or more. The flip side to this quick-turning ability is a ship that has difficulty maintaining heading without constant attention by the watch officer. By using the autopilot to reduce the workload of the watch officer, more attention can be placed on monitoring and solving the navigation picture.

Worsening weather will have a negative effect on any vessel's ability to hold course. For water jet craft this is magnified by its missing underwater structures, rudders, and keels. Seas and swells taken on the quarters can cause yawing in excess of 30 degrees while operating inside the safe-operating envelope. Hand steering in these conditions will be extremely taxing for the watch stander. Setting the autopilot controls to allow the ship the maximum of course leeway will allow the ship to average the desired course.

Seas and swells on the bow will have little effect on the vessel's ability to hold course. High-speed vessels are typically fine in the bow. This design feature is good for achieving higher speeds but results in a reduction of flotation forward. As a result there is less hull volume available to lift the bow with the swells. The effects of this will differ with the design of the ship.

Monohulls will rise fairly well with the swell but with too much speed could shoot off the crest. This will result in extreme slamming as the flatter stern hull sections land on the water surface. This type of impact may result in damaged hull plating, global damage to the vessel structure, or injury to crew members or passengers or could cause cargo lashings to fail.

Catamarans will tend not to aggressively pitch. As long as the swell does not reach up to the underside of the ship between the hulls, this can be seen as a positive. When the swell does reach the underside of the ship, often an extremely violent slam can be felt throughout the ship. This slam is called a wet-deck slam. Particularly strong wet-deck slams can cause local damage to hull plating, global damage to the vessel structure, or injury to crew members or passengers and could cause cargo lashings to fail.

Trimarans have fine bows similar to catamarans and will not pitch up with the swell, but they also have wet decks that can slam in high swells. Since trimarans are multihulls, they will largely respond to high seas in the same way as a catamaran.

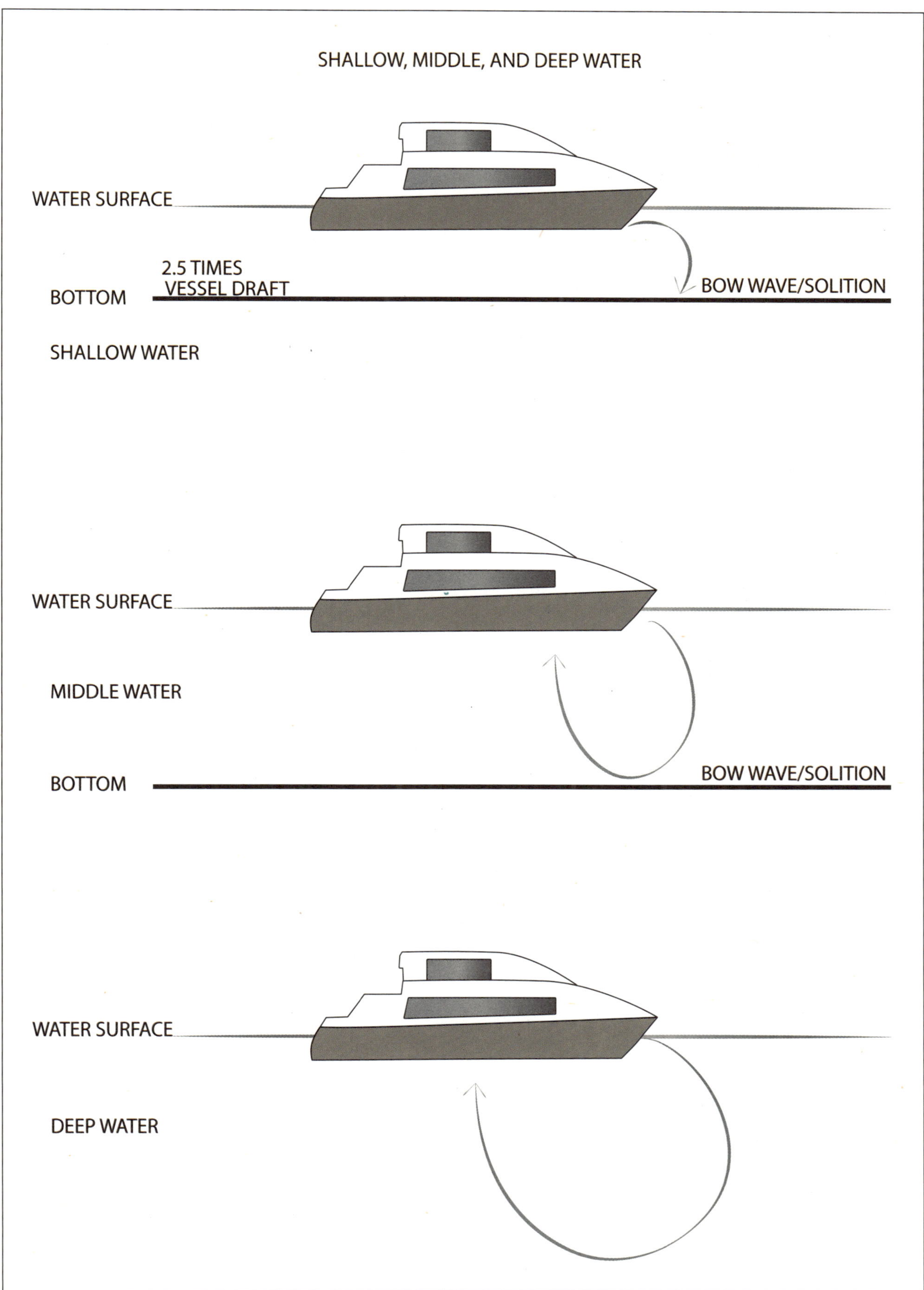

Shallow, middle, and deep water. *Courtesy of Julianne Applegate*

In extremely heavy seas, all three types of vessels will struggle. Multihulls run the risk of burying their bows in steep, long-period swells. A monohull could easily get into a synchronous rolling situation in beam seas. All three ship types may experience loss of steering if the jet inlets leave the water and the impellers run exposed to air.

Every ship will respond poorly to beam seas. Catamarans and trimarans are naturally stiff, and when they are in a beam sea, snap rolling becomes a real concern. Monohulls will fare better than other hull types. Although carrying ballast only reduces vessel speed, monohulls are more likely to be tender, opening up the possibility of hang rolls.

The most comfortable ride for any of these ships will be found by keeping the seas and swell near the stern, balancing the need to remain close to the desired heading. However, this can result in significant danger if the ship's speed is near or faster than the speed of the wave train.

The wave train can be defined as the collection of organized waves moving together. Wave trains move near the same speeds as high-speed vessels. This can present a dangerous situation if the ship is allowed to surf down the front of the swells. As the hull surges or surfs down the face of the wave, speed will go up while control goes down. This is another way the thrust-to-steering ratio can be negative. Once at the bottom of the swell, the bow may dig in and cause broaching, or bow submergence. Vessels with fine bows are more likely to submerge the bow. Multihulled vessels are more likely to broach significantly.

STEERING MODES

Various steering modes may be available depending on the manufacturer of the steering-control system and the individual layout of each ship. Typically it is possible to control the jets independently. Additionally, there will normally be a mode that allows all the jets to be controlled together. When operating in a channel or out in open water, it is recommended to have all the jets working together. This allows for simplified control of the ship and maximum effective steering.

STEERING IN A CHANNEL

Piloting a high-speed vessel down a narrow channel is very similar to any other vessel. However, there are some differences in how high-speed vessels interact with the bottom and how vectored thrust actually steers the ship.

BOTTOM EFFECTS ON LOW-DRAG HULLS

In shallow water—commonly understood to be less than 2.5 times the draft of a given ship—low-drag hulls will behave very similarly to traditional displacement hulls. They experience the same increased stopping distance, increased turning diameters, reluctance to respond to helm commands, and tendency to squat as speed is increased or underkeel clearance is reduced.

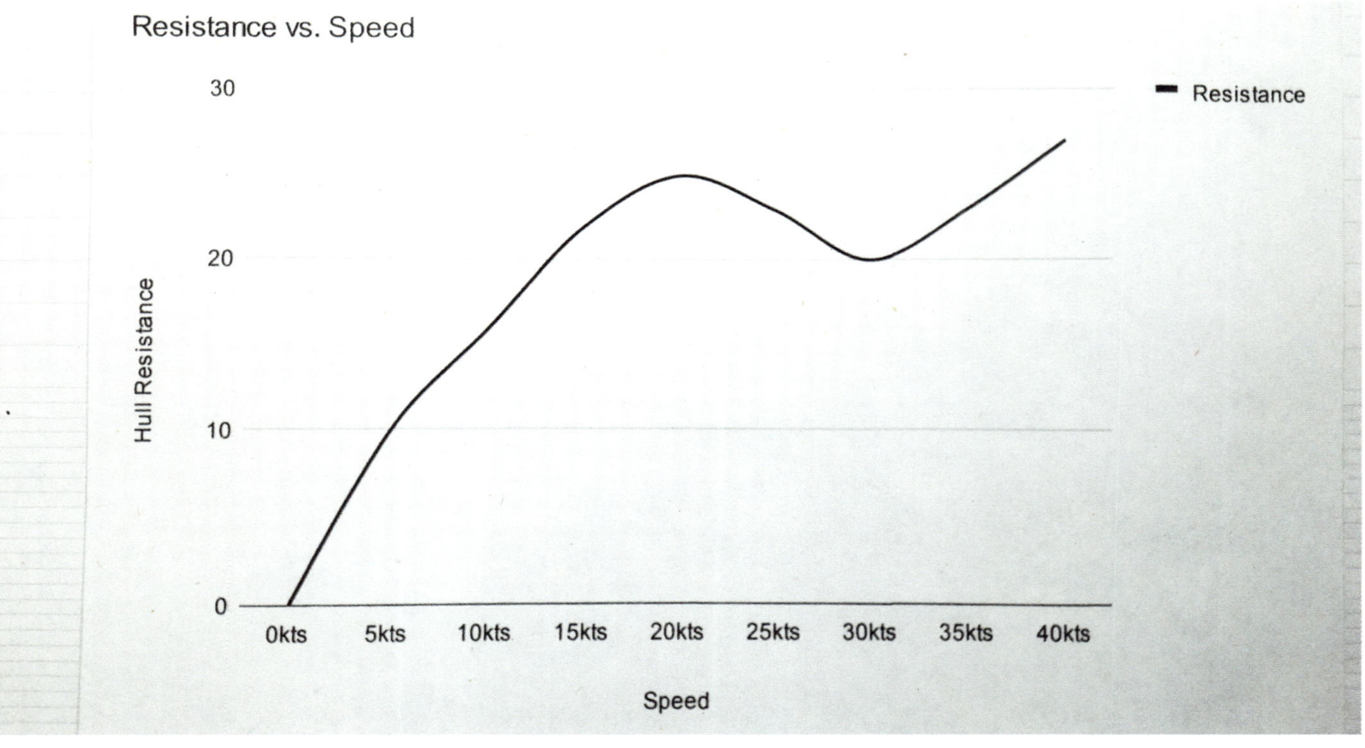

Generic graph showing hull resistance compared to speed. Note that at a certain speed, resistance significantly reduces. The highest point of resistance is the vessel's hump speed. Once the resistance reduces, engine power can be reduced while speed will remain the same, since the vessel is on plane and traveling with greater efficiency. *Courtesy of Julianne Applegate*

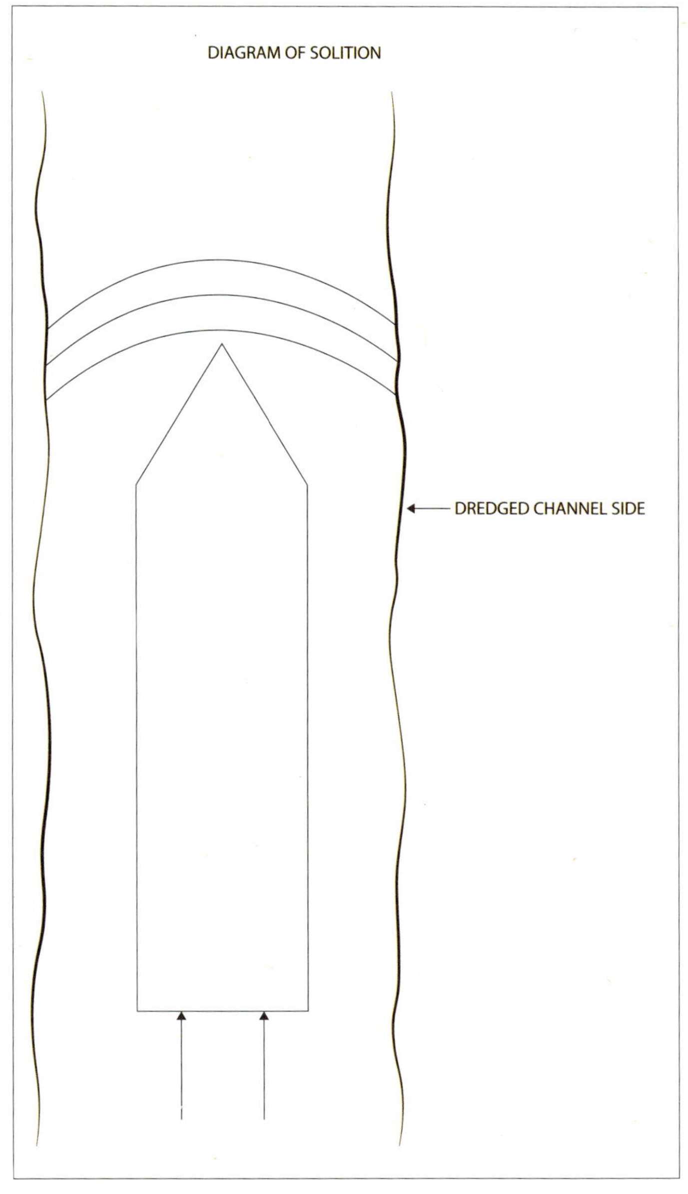

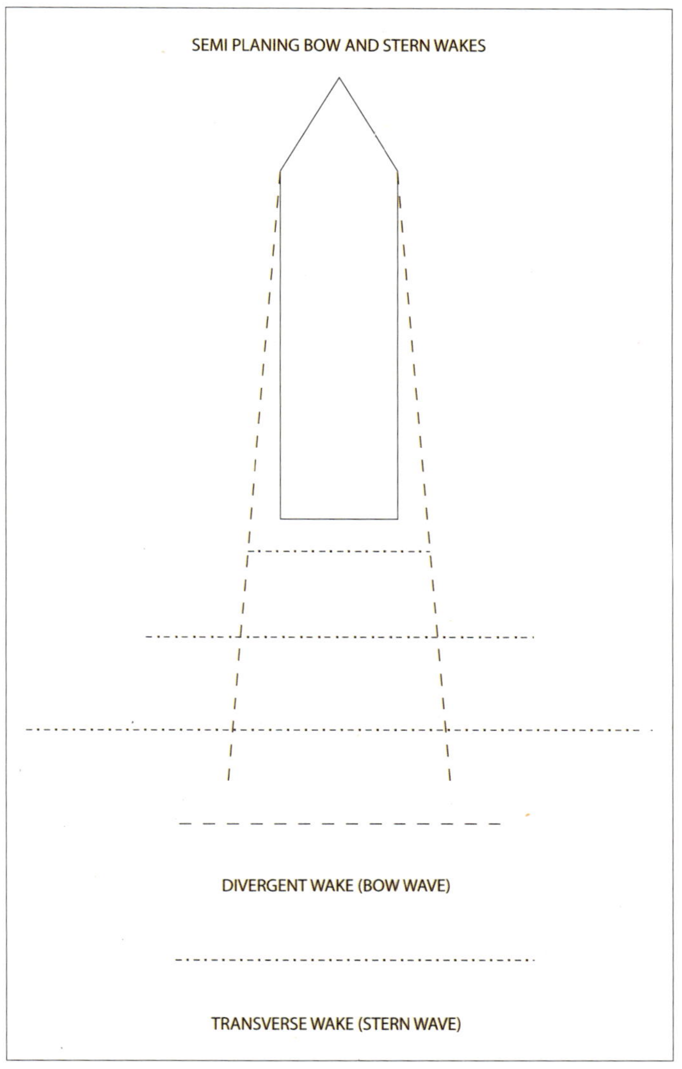

Left: Soliton outpacing ship in a narrow channel. *Courtesy of Julianne Applegate*

Above: Semiplaning-hull wakes. *Courtesy of Julianne Applegate*

The water depths between what is commonly understood as shallow water and deep water deserve more discussion. This "middle water" is best defined as water depths that are more than 2.5 times the draft of the vessel, but not deep enough to allow the bow wave to come back under the ship to lift it up on plane without touching the bottom. This depth varies largely with vessel length and weight. It is in these water depths that high-speed vessels can get into significant trouble. A high-speed vessel is able to get up on plane in middle water, although the power required and the wakes that can be generated are quite high. The water depth required for the bow wave to come back under the ship in this way will vary depending on vessel displacement and waterline length.

Assuming a ship has the power to squeeze the bow wave under the ship, planing is possible. However, there is the potential for increased wake and soliton creation and a higher likelihood of cavitation. A vessel unable to squeeze the bow wave is at a higher risk of developing thrust breakdown. Keeping the vessel below planing speed while in middle-water depths will prevent the creation of potentially large wakes and solitons and reduce the likelihood of cavitation. If sufficient distance exists between the vessel's tack and areas of potential damage—that is, piers, breakwaters, beaches, anchored vessels—operating at speed in middle water may be permissible on a case-by-case basis.

If the bow wave is squeezed beneath the hull and cavitation is not happening, efficiency and speed will be higher than when operating in deep water. Cavitation is more likely in the middle depth zone due to the bow wave pushing water out of the way.

When going from deep water to middle or even shallow water, speed and efficiency can be expected to go up, along with the size of the wake. The increase in speed and efficiency may be temporary for heavier or larger vessels. While drag on the hull is reduced, the ship still has to drag the bow wave along, and this is interacting with the bottom in middle and shallow water. If sufficient power is not available to drag the bow wave, the ship will fall off plane.

INCREASING AND REDUCING SPEED

Increasing thrust will increase the speed of the vessel, provided that thrust breakdown or adverse bottom effects are not present. Increased thrust will also improve steering, with the same limitations.

Reducing thrust will decrease steering control. A significant reduction in steering control can be associated with large reductions in thrust. It is possible to completely lose all steering control when going from a high thrust speed level to a low level. As mentioned, this is caused by a negative ratio of thrust available to steer the vessel compared to the speed of the water flowing past the hull. Steering will remain degraded until the ratio of thrust to water speed is returned to a positive level. This can be achieved by either increasing the level of thrust or allowing the vessel to slow.

HUMP SPEED, SQUAT, AND CAVITATION

Hump speed is defined as the speed at which a vessel is able to "climb" onto the back of its own bow wave. This is also known as being on plane. A vessel's hump speed will vary depending on the waterline length, displacement, underwater hull shape, and the use of ride control surfaces.

The most efficient speed will often be just higher than hump speed. While a great deal of power and fuel may be required to get a vessel over the hump, once on plane, power can be reduced below the level required to get over the hump. This increase in efficiency is a result of the change of trim that occurs as a ship comes fully up on plane, and from the fact that the ship is now "surfing" on her own bow wave. The water supporting the ship is already moving forward, much like a surfboard on a breaking wave.

The causes and results of squat are the same both for displacement and high-speed vessels, but the effect is magnified on water jet vessels. A significant low-pressure area develops along the after end of the hull near the water jet intakes, due to the increased speed of the water flowing under the hull and the suction of the water jets running above idle rpm. This decrease in pressure at the aft end of the ship causes an increase in squat, and the angle of attack the waterline has with the surrounding water. Ride control surfaces, trim tabs, or interceptors are installed to make the process of getting on plane more efficient and have the added effect of reducing the likelihood of cavitation. However, even with ride controls installed, cavitation is still possible if pump rpm outpaces vessel acceleration.

WAKE MANAGEMENT

All vessels produce wakes. When high-speed vessels produce wakes, they are compounded by the amount of power driving the vessel forward and the shape of the actual hull. There are individual waves that come off the hull at different places.

The soliton wave is a type of wave produced by all ships moving through the water. This wave is a function of the bow wave. As a ship moves forward, water is pushed out to the sides; additionally, water is pushed forward. The water being pushed forward creates the soliton. In shallow and middle water, the bow wave is not able to rotate under the hull. This wave instead reaches down to the bottom of the channel. This water will form a standing wave at the bow. As a ship increases speed, this wave will grow and a narrow channel will magnify the level to which this wave grows. Once this wave is created, there is no way for the ship to stop it. Slowing down will only cause the soliton to proceed ahead of the vessel until the wave reaches a deeper or wider stretch of water.

Another vessel operating in a confined channel could be pushed out of the channel by this wave. The soliton can produce conditions similar to a following current, reducing the steering effectiveness of a vessel being overtaken. A vessel alongside a pier may part her mooring lines after being subjected to such a wave. Vessels at anchor may find that their anchors drag while under the influence of a soliton. There is also the potential for large breaking waves if a soliton were to pass over an area of shallow water.

The wake that occurs out to the sides of a ship moving forward is referred to as the divergent wake. These wakes form to either side of the stern and are more commonly called bow waves. As speed is increased, the angle that these wakes make with the hull will be reduced until the wakes are virtually parallel with the hull.

Aft of every ship is the standing stern wave, properly defined as the transverse wake. This wave is created by the water that has been squeezed under the hull and is finally being able to come back to the surface. As speed increases, the size of this wave will also increase. Planing-hull vessels will see this wave as part of their rooster tail.

When the divergent wake and the transverse wake finally meet, they interact in a way that causes the resulting wake to have an increased height and power. How much increase in height and power is a function of the angle of intersection. The closer to right angles the wakes get, the more power and height that can be created from their meeting. It is important to understand that above hump speed, the divergent wake lies at a shallower angle to the hull, resulting in a wake intersection angle closer to 90 degrees. The resulting wake typically forms three nodes, or peaks, that can travel a number of miles out to the side of the ship's course, causing damage long after the ship that created the wakes has passed.

NARROW CHANNELS

Steering a water jet vessel in a narrow channel will be fairly similar to a traditional ship. Bucket angles and the frequency of change are tied to the ratio of thrust to speed. This is very manageable and not an issue of great concern.

Like any other ship, water jet ships may have to proceed down narrow channels with high environmental forces. For propeller and rudder ships, this often means transiting at a high speed, crabbing at an angle to the channel, or slowing down—typically less than 3 knots—to a point where a tug can come alongside and hold the ship in the channel. Water jet ships are able to use thrust vectoring to hold the ship against these forces while heading down the channel at reasonable speeds. Proper use of thrust vectoring results in a ship with a bare steering speed that can very closely approach zero knots.

Narrow channels in and of themselves require a reduction in speed. At some point, reducing speed will result in the reversing bucket partially closing from the full open position. When this happens, effective thrust is significantly reduced. Attempting to steer a ship with partially closed buckets will result in a bare-steerage situation. To avoid this condition the buckets should be split. (The methods of splitting the power plant are discussed in more detail in the section on maneuvering.) The buckets on the windward side of the ship are placed to the astern position so that the buckets are completely closed, providing 100 percent astern trust. This will allow the buckets on the leeward side of the ship to remain 100 percent open, maximizing ahead thrust. This ahead thrust is used to steer the ship. Increasing the astern-side water jet rpm will allow for greater control of the ship but will result in a slower top speed. It is important to note that increasing the rpm of the astern jets above the level recommended at ZSTW could result in a cavitation condition and should be avoided. This jet arrangement is referred to as split plant.

An additional effect of splitting the plant is that the vessel will naturally twist toward the side of the backing jet. Placing the windward jets astern will help pull the bow into the wind. Careful thrust vectoring will allow the vessel to proceed along the channel without crabbing or developing leeway.

Environmental forces will eventually overcome the ability of split plant to hold the center of the channel. At this point, crabbing may be required, although the required angle of crab will be much less than that of a ship with propellers and rudders.

Many channels will require meeting other vessels. While most channels are wide enough for safe passage, there are channels similar to the Houston Ship Channel where vessels must meet with very close CPAs (closest point of approach), often measured in feet instead of tenths of a mile. Typically in channels this tight, a maneuver known as the Texas chicken is used to keep the meeting vessels from colliding.

The Texas chicken makes use of and allows the ship's officer to deal with the various positive- and negative-pressure areas that surround a ship making its way through the water. The two ships start out with a fine angle on the bow, heading straight for each other. Speed is not slacked prior to meeting, since this would reduce the positive-pressure area present at the bow. At a distance relative to the expected amount of pressure at the bow, the ships turn slightly toward the edge of the channel. Just before the bows are even, the rudder is put toward the other vessel, preventing the soliton from pushing one of the ships out of the channel and aground. As the bows pass each other, the rudder is centered to hold the ship straight as they pass. Before the bow is even with the stern of the other ship, the rudder is put away to keep the bow from being sucked into the negative-pressure area present at the stern of every ship. As the ships pass each other's sterns, the rudder is eased, then put over toward the center of the channel to hold the sterns away from each other. Once the ships are clear of each other, they continue to steer for the middle of the channel as before. This is a complicated and delicate maneuver that is normally undertaken by pilots experienced with the channel, local customs, and the various weather conditions. As channel width increases, the effect of the intership forces will decrease, as well as the need for this maneuver.

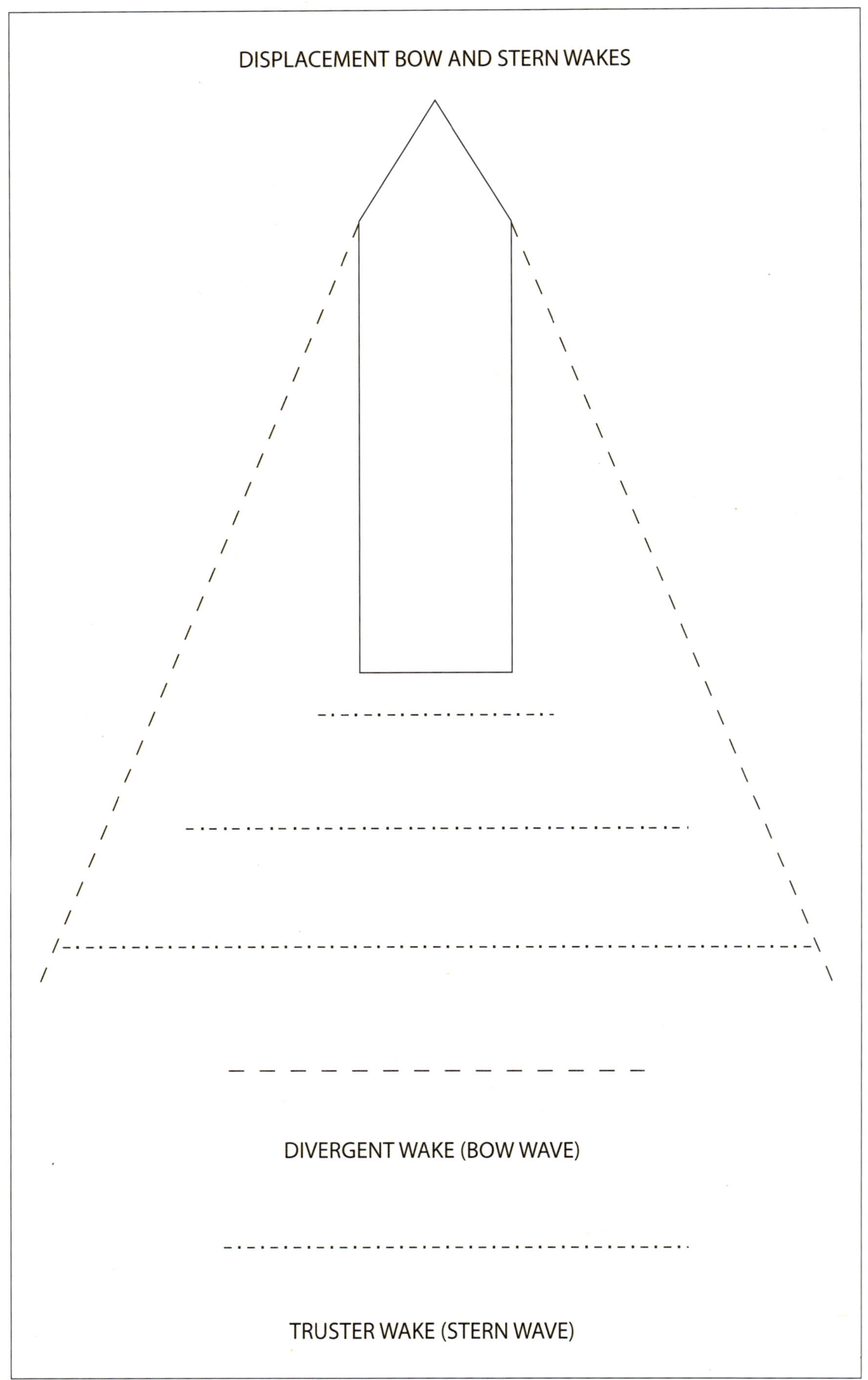

Displacement-hull wakes. *Courtesy of Julianne Applegate*

For a high-speed water jet ship this maneuver is even more complicated, since the various high- and low-pressure areas form at different spots along the hull. When on plane, the high-pressure area forward is barely noticeable, whereas the negative-pressure area directly astern is quite high inside the wake. The wake itself could present more like a high-pressure area instead of the expected low-pressure area. With this in mind, close-quarters situations with other vessels, even if local custom allows them, should be avoided while on plane.

Special care is required when coming up on plane when meeting other vessels. The angle of attack that the hull has with the water is at its most extreme at this point and may cause an unexpected hull pressure interaction. It is safer to wait for any close-aboard vessels to get well clear prior to accelerating.

The suction of the jets at the stern will cause a further reduction in pressure at the stern, largely depending on the distance the intakes are from the extreme sides of the hull. This may cause a larger-than-expected suction on the hull of the other vessel. Unfortunately, increasing rpm to get clear of the other vessel may only make things worse. At this point, while counterintuitive, it would be better to reduce rpm to allow the suction to ease. Many of these problems present themselves when overtaking in narrow channels.

Many water jet vessels are primarily envisioned to travel at high speeds, requiring them to be light for their overall size. This will cause them to get pushed and sucked around by these inter-ship forces more than other vessels of comparable dimensions.

To complete a tight turn while remaining at a slow speed requires "pumping the rudder" with a traditional prop and rudder vessel. While water jet ships are highly maneuverable, there may be times when a similar technique is useful. Here water jets can be "pumped" the same as other ships, but if the rpm is left high for too long, the water jet ship will increase not only speed, but the diameter of the turning circle as well.

All these issues should be included in master pilot conferences. Having a pilot take the conn can be problematic. Few pilots consistently conduct enough trips on water jet vessels to remain current on the differences in handling compared to a displacement ship. It is far better to work with pilots, making use of their specific port knowledge while clearly explaining the transit and maneuvering plan. When involved in this way, pilots are more willing to provide feedback, rather than commands that may not match the actual maneuvering abilities of the ship.

CHAPTER THREE

HANDLING ALONGSIDE

The thrust-vectoring nature of water jet ships allows for superior handling while alongside a pier. There are various ways of conceptualizing the process of getting these vessels to "walk" and spin though the mooring evolution. It is important to understand that no matter how the ship handler visualizes the control and movement of the ship, it all comes down to vector addition and how control inputs change the way the resultant vector affects the vessel's movement.

In practical terms, the only motion that matters to the shiphandler is the relative motion of the ship compared to the pier being landed on or departed from. Being dead in the water (DIW) is not the same as being stopped holding position against the various forces of wind and tide as they push on the vessel.

The preferred and most accurate position on any bridge to observe relative motion while docking is the bridge wing. Most vessels have jet controls, indictors, and navigation instruments on the bridge wings for this reason. If the bridge wing does not have good navigation and bucket position indicators, the value of this position is diminished.

There are some water jet ships where the only position to dock the ship is amidships on the bridge. These stations typically will have CCTV screens positioned so that the docking officer can see the sides of the ship while operating the controls. These cameras do not replace the need for eyes on the side of the ship. A trusted lookout who knows how to handle the ship will need to be positioned where they have a good view and an easy means of communication back to the docking station.

A center docking station will most likely be able to see the various navigation information systems. These systems are not a replacement for eyes on the side of the ship. While an ECDIS or speed log might seem to present accurate data regarding the movement of the vessel, they are rarely if ever current with the picture out of the window. GPS receivers update about every two seconds. Speed log errors could be observed if reverse wash from the backing jet makes it to the sensor position; still, these sensors are measuring a change in speed that lends itself to a delay. While two seconds may seem like a short duration of time to be worried about, water jet ships are capable of going from 0.5 knots astern to 0.5 knots ahead in less than the two seconds it would take the ECDIS to update. Changes in heading are similarly quick.

When observing relative motion, it can be difficult to discern whether the vessel is twisting (rotating) or walking (sliding sideways). The best way to tell the difference is to look for the point in the window where there is no relative bearing drift to the right or the left. This is the direction that the ship is going—the direction of the ship's vector. If this point cannot be found, the ship is twisting. If this point is found off the beam, the ship is walking.

The lack of a fixed point in the window makes it difficult to determine where the vessel is setting while twisting. For this reason it is recommended that twists be conducted well away from any potential dangers, when possible.

The best way to keep the point of no relative bearing change steady is to be very focused on the heading angle of the slip or pier face. Staying on heading will help simplify the pier work process. In fact, it can be said that relative motion makes more sense when staying on this heading. This will help guarantee a flat landing along the pier, which is imperative with lightly built high-speed vessels.

The other visual cue that should be closely watched is the position along the pier. If possible, find a natural range to line up on once the vessel is in position.

Maintaining constant attention on these two points of information will allow the shiphandler to have a good feel of the relative motion of the ship. Other sources of information—ECDIS, speed logs, ROT indicators, radar—are also useful, but these should be looked at as more of a reality check. The situation out of the window must always remain paramount.

EFFECTS OF ONE JET

To understand how to make the most use of the vectored thrust that is available with water jet vessels, it is first important to understand how each jet is affecting the ship. Maneuvering a single jet vessel is very similar to working an outboard or stern-drive vessel. The largest difference is that the water jet vessel is always pumping water.

To keep the vessel from moving ahead or astern, the bucket is set to the "Zero" position. This will allow approximately 60 percent of the wash to be diverted astern, leaving the remaining 40 percent ahead. An even 50-50 split in the wash will cause slight ahead motion, since there are losses in efficiency when the wash is diverted astern.

If the water jet is set in the ahead direction, the boat will move ahead. The amount of ahead speed is controlled by how much bucket is opened, and once fully opened how much rpm increase there is. Heading is then controlled by the angle the nozzle deflects the wash to: port or starboard. Just like with an outboard or stern drive, the jet is angled toward the side that the bow needs to go to (e.g., starboard bucket for a starboard turn). Again, similar to the outboard or stern drive operating astern, the stern will go in the direction the jet is angled toward. An astern jet to port pulls the stern to port, and the bow goes to starboard. Another way of thinking about this is to imagine a tug made up to the stern in the same place as the jet. A tug pushing ahead, but at an angle from the keel, will cause forward movement while turning the ship toward the side it is angled to. The same tug pulling aft on the stern at an angle will pull the stern in that direction.

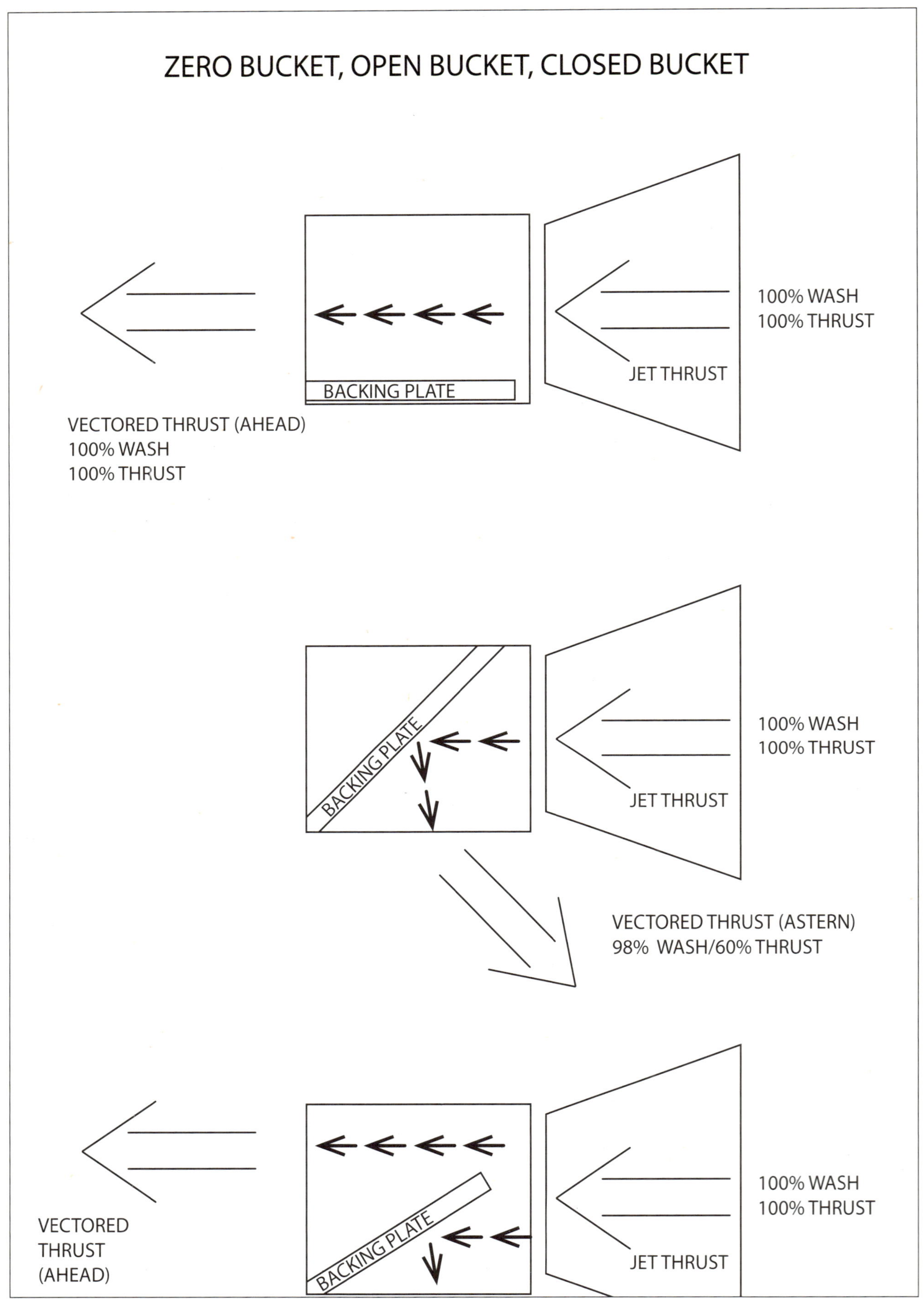

Zero bucket. *Courtesy of Julianne Applegate*

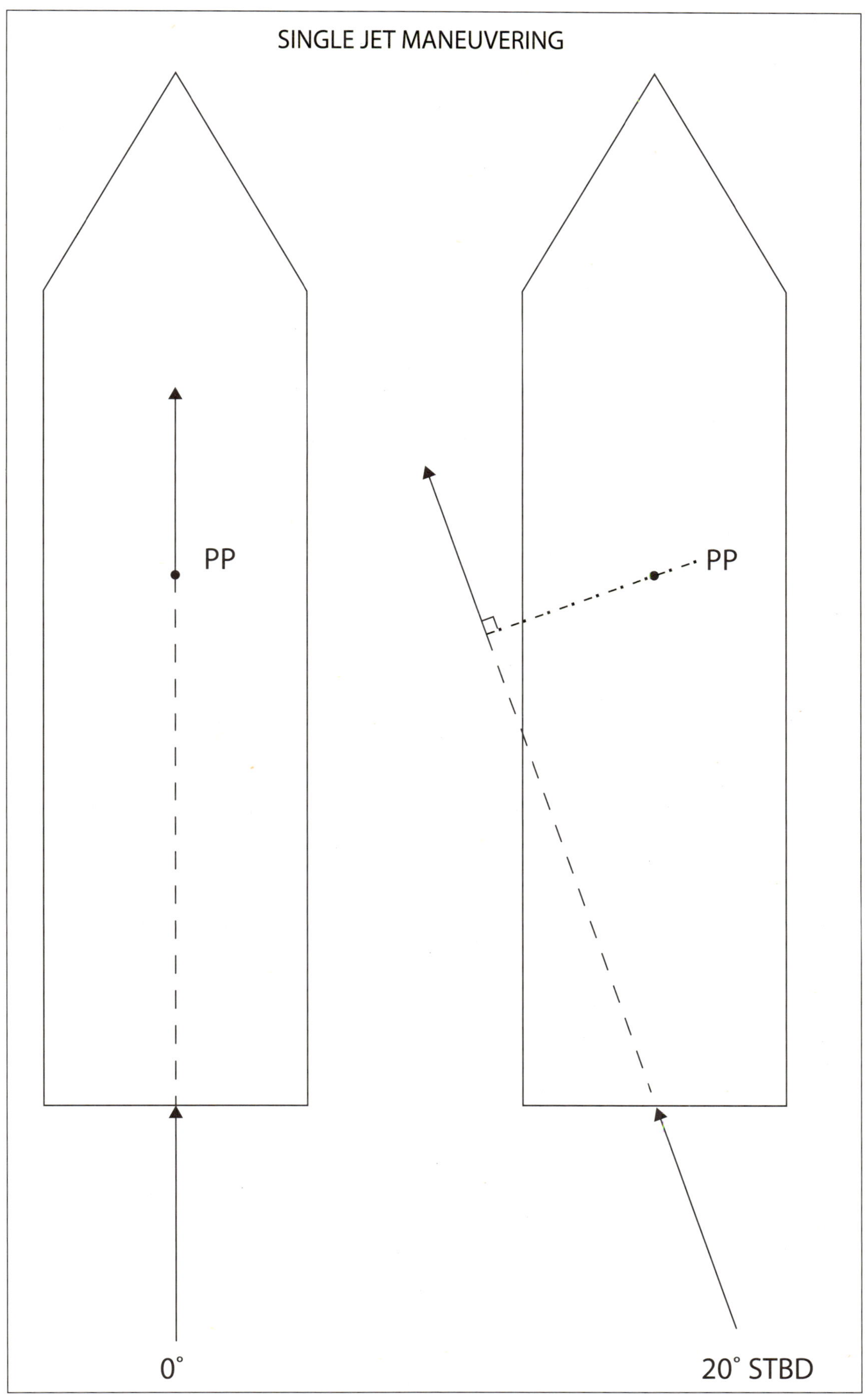

These drawings show how a force vector applied through or away from the pivot point will cause the ship to go straight or create a lever arm to turn the ship. *Courtesy of Julieanne Applegate*

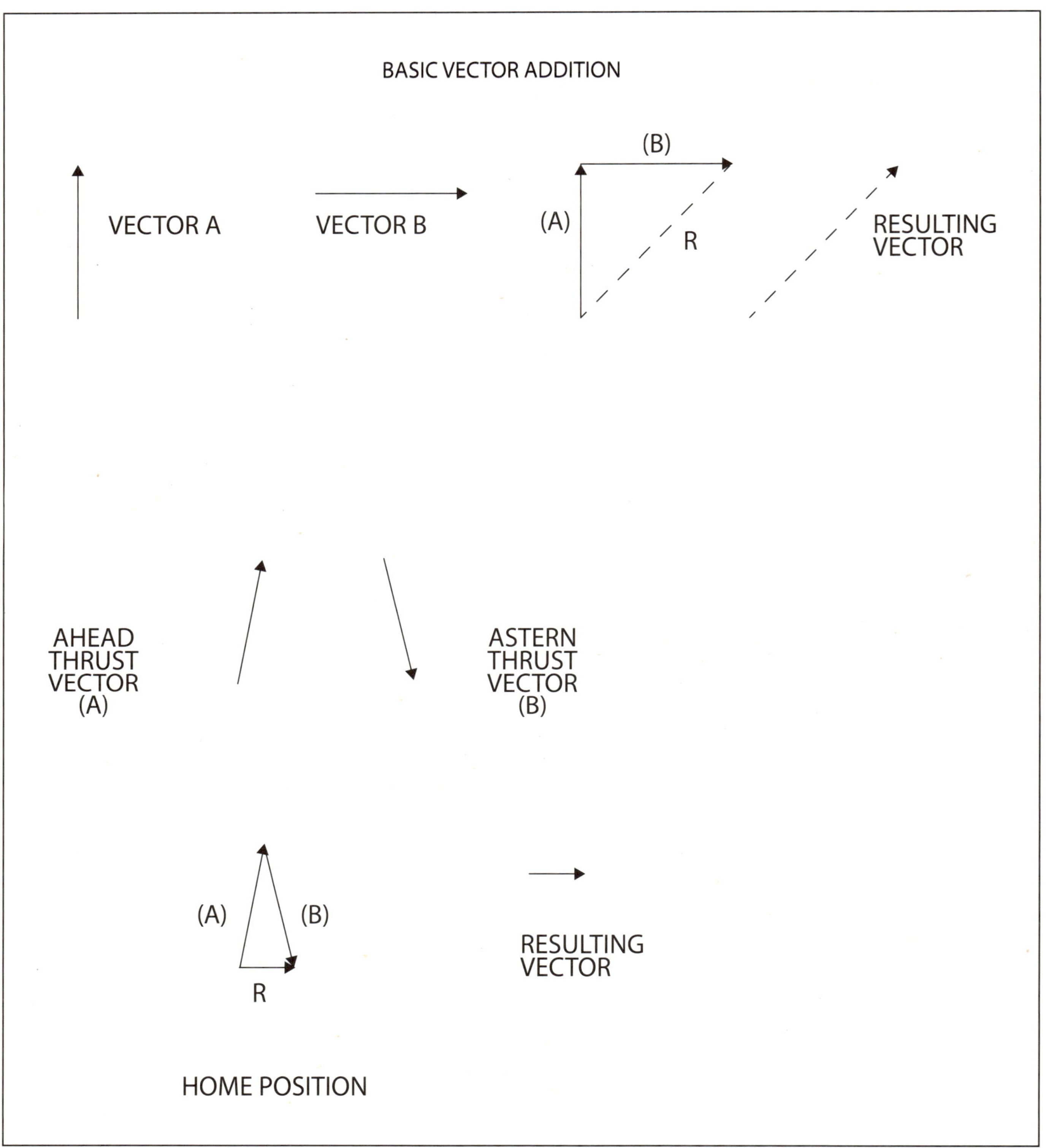

Two examples of vector addition. The second example shows the possible home position of a water jet ship. *Courtesy of Julianne Applegate*

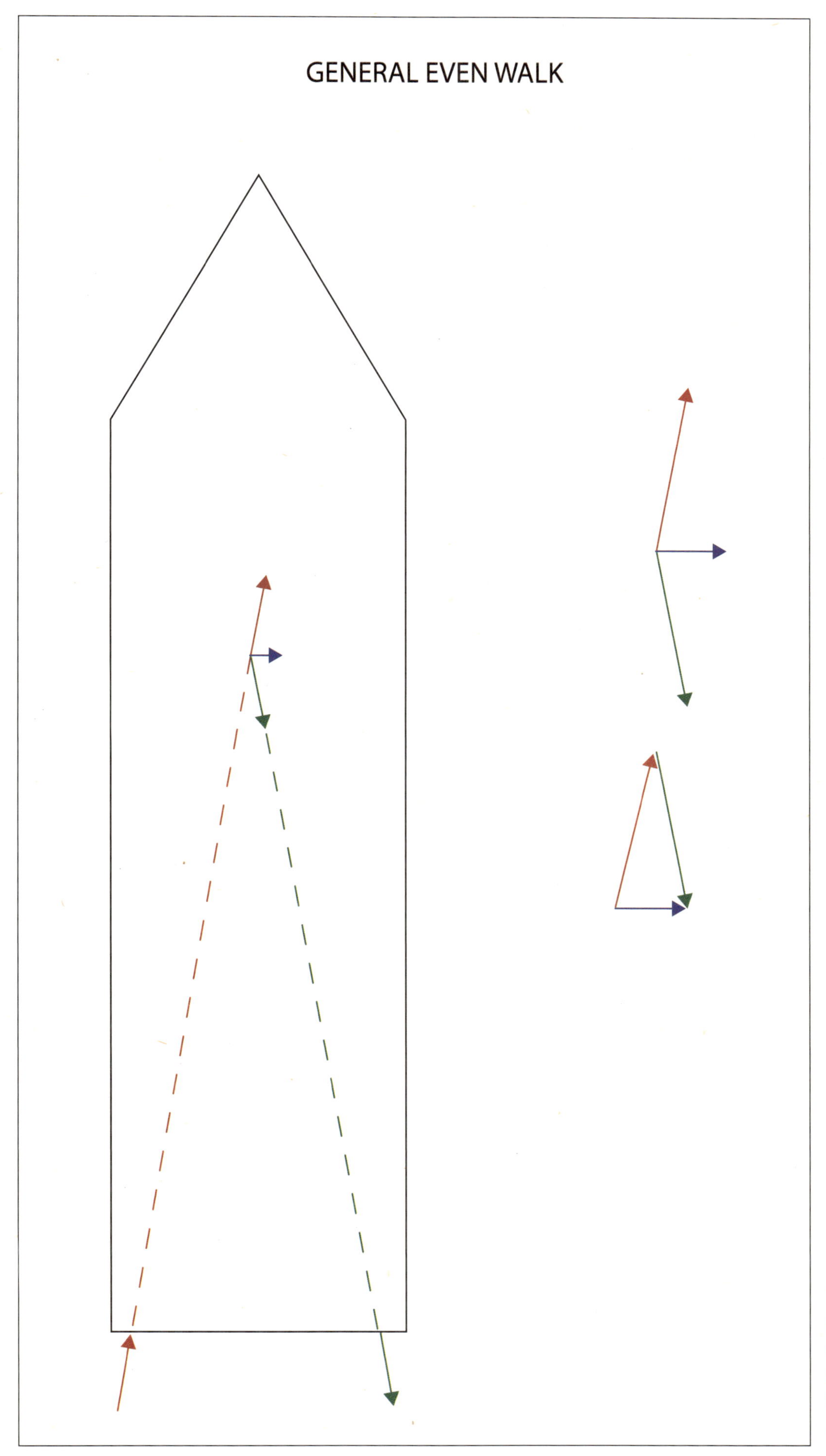

Two examples of vector addition. The second example more correctly shows the home position for this ship. *Courtesy of Julianne Applegate*

MULTIPLE JETS

Jets offset from the keel will tend to cause the vessel to turn away from the side the jet is on. A ship proceeding only on the starboard jet will tend to continue turning to port with the jet centered unless corrective action is taken. This is true for any vessel equipped with more than one engine or jet. One large difference here is that at small angles, the ship will not so much turn as it will crab. The vectored nature of the thrust pushing through the pivot point causes little to no turning lever arm.

The farther the jets are offset from the keel, the larger the lever arm that develops between the pivot point and the extended vector line.

SPLIT PLANT

To maintain zero speed through the water with multiple jets, it is best to split the direction of the jets instead of trying to keep all the jets at the zero position. Errors in jet control calibration make this almost impossible. The astern wash is not as efficient as the ahead wash due to power lost at the backing plate, and the angle of the wash under the hull. This requires that astern rpm be higher than ahead rpm to balance the power, resulting in no net thrust. How much higher will vary ship to ship, even among sister ships. Generally the ahead wash needs to be about two-thirds as powerful as the astern wash.

Steering the ship in split plant requires an awareness of the power bias. An ahead bias in the split will cause the ship to move forward. An astern bias will move the ship aft. The bias is controlled with the power of the ahead jet. An increase in ahead rpm above even power results in an ahead bias. Reducing ahead rpm will cause an astern bias.

When the bias is even or ahead, the ahead jet should be the active jet for steering, since there is more effective flow passing through the bucket for thrust vectoring. When the bias is astern, the astern jet should be the active jet. In this condition the flow passing through the ahead bucket has been reduced below the point at which steering is more effective than the astern flow.

The choice of which side of the ship should have the astern jet is largely determined by environmental forces. To provide the greatest control of the vessel, the backing jet should be on the side of the strongest environmental force. This enables the lever arms to work most efficiently.

PIVOT POINT AND INTERSECTION POINT

The pivot point on a water jet vessel will move in the exact same way as the pivot point on any other vessel. Increase speed forward, the pivot point will move forward. Go astern, the pivot point moves aft. Turn aggressively, and there is no change until the overall speed through the water changes. Set the engines against each other to twist, and the pivot point stays approximately in the center of the ship.

As mentioned, vessels with multiple jets can split the plant to allow for exceptional maneuverability. By balancing the force of the ahead wash against the astern wash, the ship can be said to hover, or have zero net speed through the water. However, even with the jets midshipped, there are lever arms working on the vessel. They will tend to cause the ship to twist the same way as a propeller-driven vessel.

Rudder angle is typically referenced to the keel, as in port 10 for a rudder that is 10 degrees to port from the keel. This works for rudders, since there is normally only one, or multiple rudders are linked together. Water jet boats with simple steering systems will link the buckets together. These vessels are driven more like traditional ships, although control is still much higher, and it is possible to walk these types of vessels sideways. The power of the walk will be reduced compared to a ship with independent buckets.

Ships with independent buckets use different terminology to describe the position of the buckets. Buckets are referred to as toed in or out. This prevents confusion while describing the port bucket angled 10 degrees to starboard and the starboard bucket angled 10 degrees to starboard. Instead, these same buckets would be described as the port bucket toed in 10 degrees and the starboard bucket toed out 10 degrees. The direction of toe is referenced to the keel. Buckets that point toward the keel are called toed in. Buckets pointed away from the keel are toed out.

When the angles of the jets are equally toed in or out, the vectors combine at the point where the extended vector lines cross. Assuming that the lines of the vectors could be infinitely extended, regardless of the actual direction or amount of force, they cross. This crossing point is defined as the intersection point.

The intersection point is used to describe and simplify how various angle and power settings will affect the ship. This point should be envisioned as the end of the lever arm acting on the pivot point. It is envisioned because all the force generated by the jets is actually imparted at the transom.

The position of the intersection point compared to the pivot point sets up a lever arm. The direction and force of the resultant vector act to move the ship about, whether in a rotational (twisting) or transitional (walking) way. When the intersection point is in the same spot as the pivot point, there is no lever arm and therefore no rotation, resulting in lateral motion. As the two points separate, the lever arm grows. The amount of rotation is a function both of the strength of the resultant vector and the length of the lever arm.

For very small angles the intersection point is well outside the skin of the ship, producing an incredibly long lever arm. While the length of this lever arm is quite long, the angle that the vectors make at the intersection point is small. The resulting vector is very short, producing a minimal amount of twisting force.

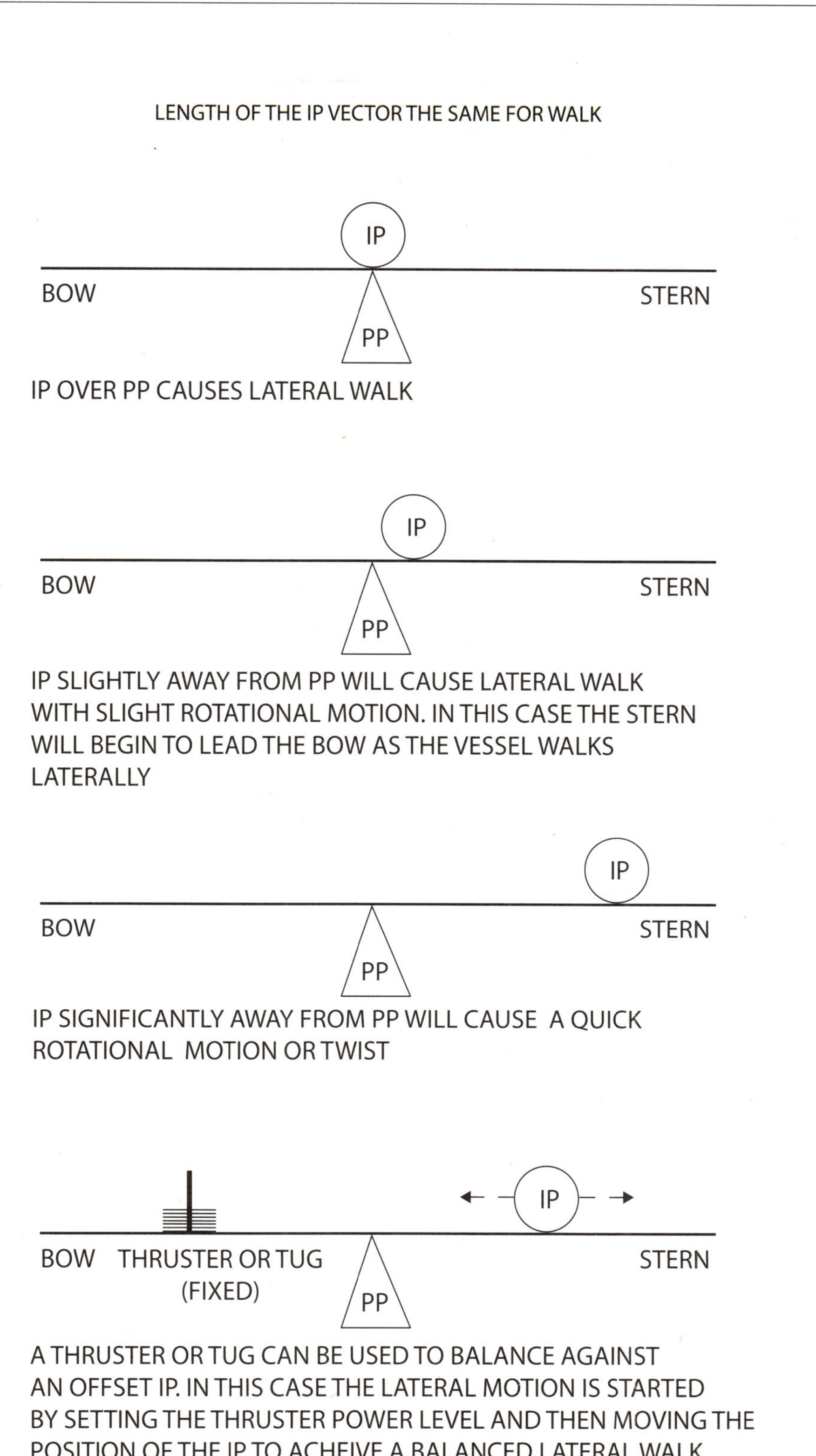

In order to walk the ship, the intersection point needs to be kept close to the pivot point. When the intersection point moves too far from the pivot point, the ship will no longer walk, twisting instead. However, if a tug or thruster is used to balance an off-center intersection point, a balanced walk can be achieved. *Courtesy of Julianne Applegate*

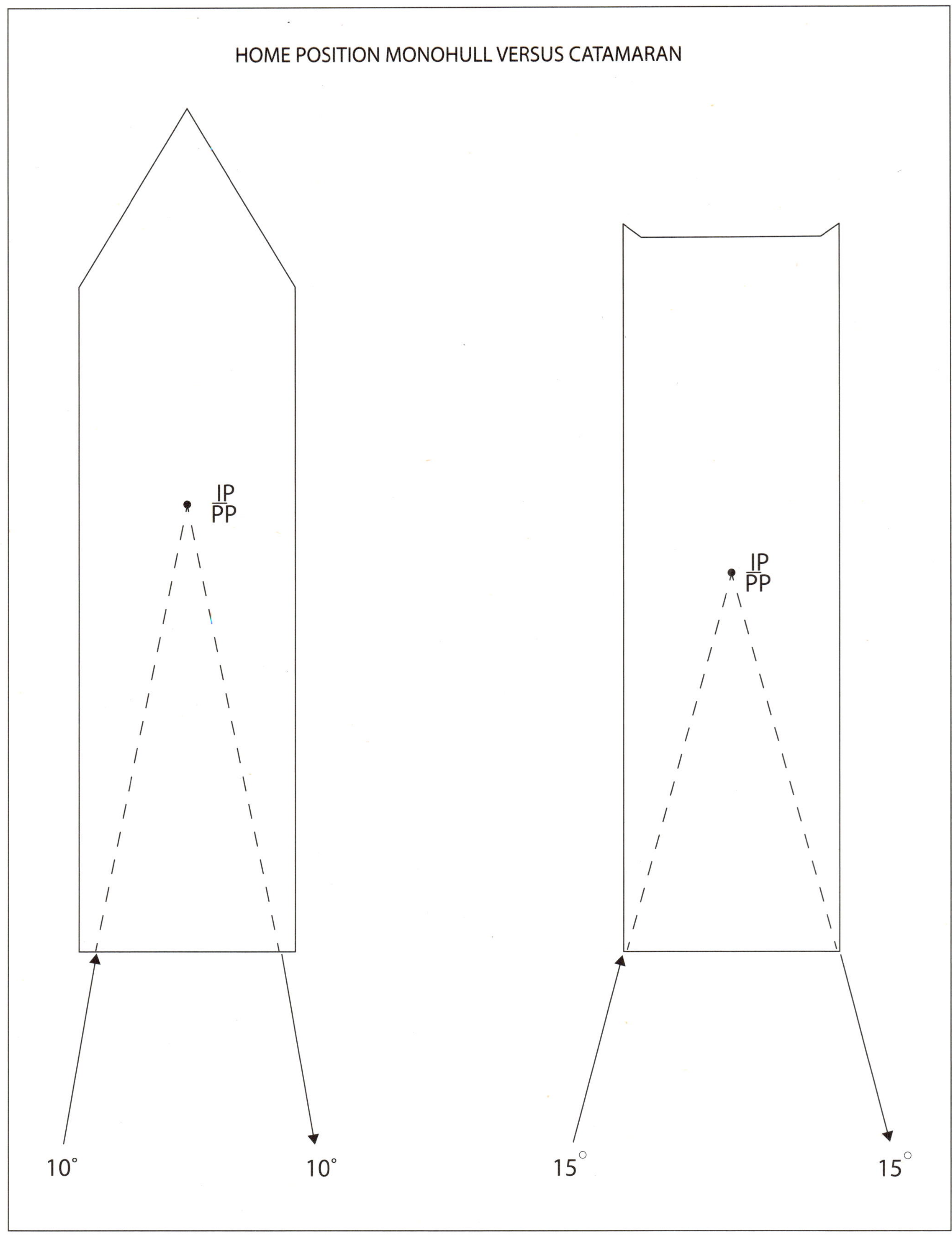

Catamaran versus monohull home positions. The catamaran home angles are much larger than the monohull due to increased offset distance compared to length. *Courtesy of Julianne Applegate*

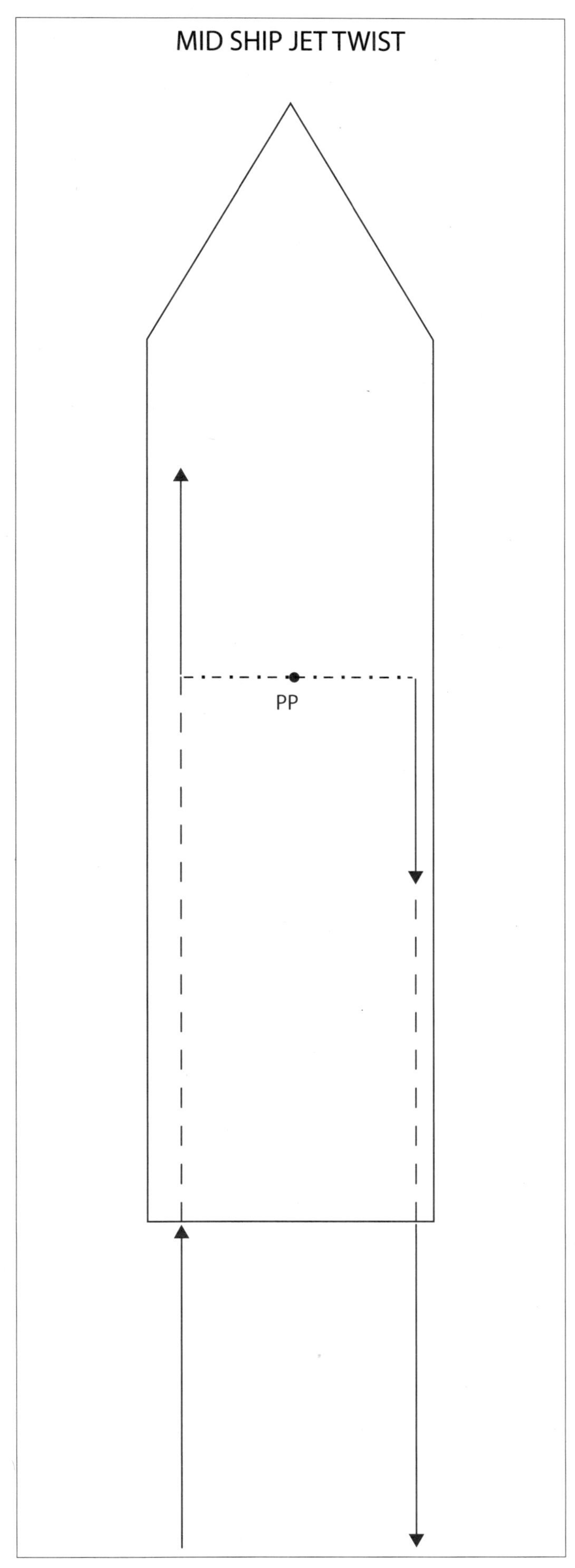
MID SHIP JET TWIST
PP

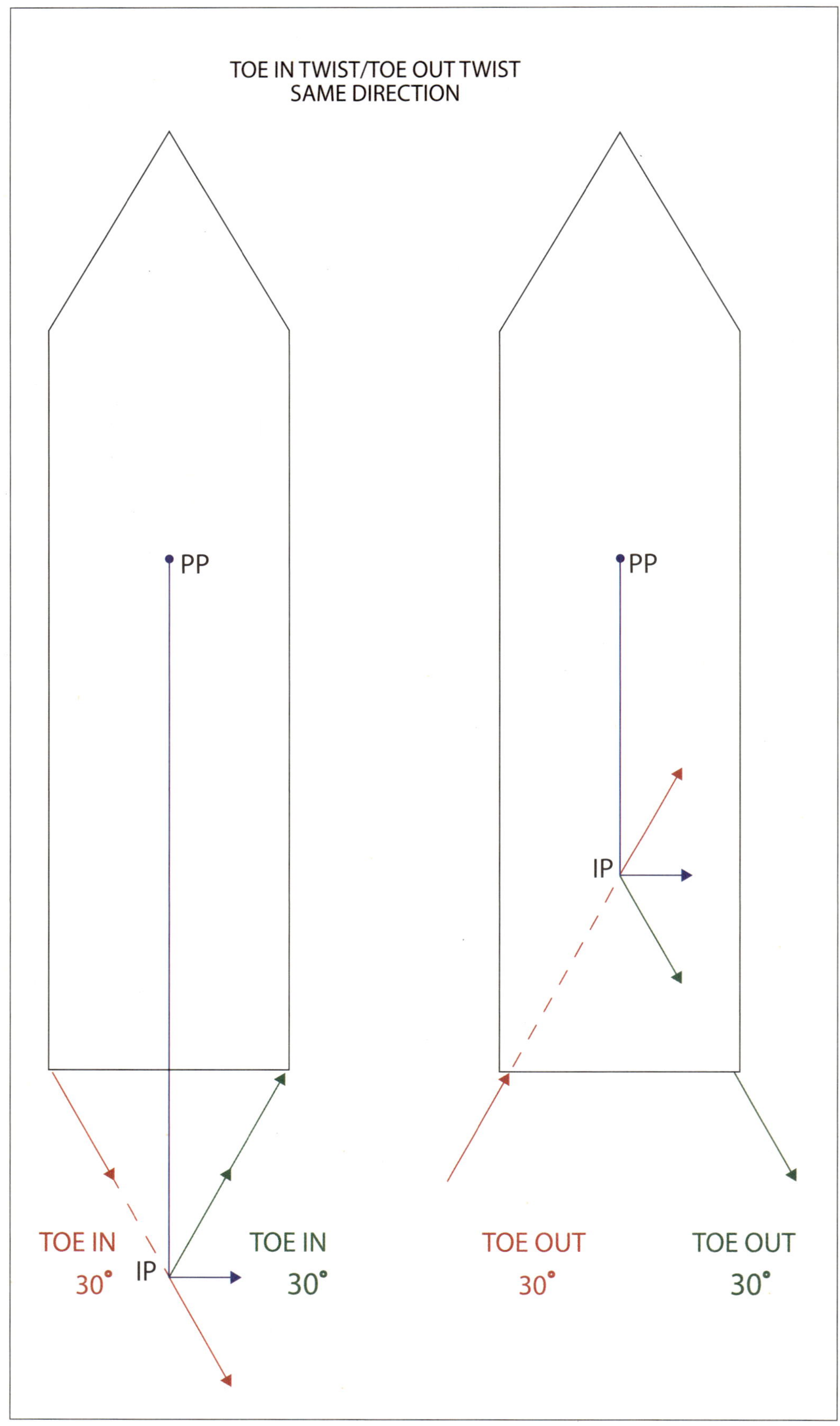

Two types of high-thrust-angle twists where the vector lines cross, greatly combining and increasing the amount of force that can be applied and resulting in a very quick twisting motion. Note: the vector line are reversed in the two twists, but the direction of twist is the same. Toes-in spin / toes-out spin. *Courtesy of Julianne Applegate*

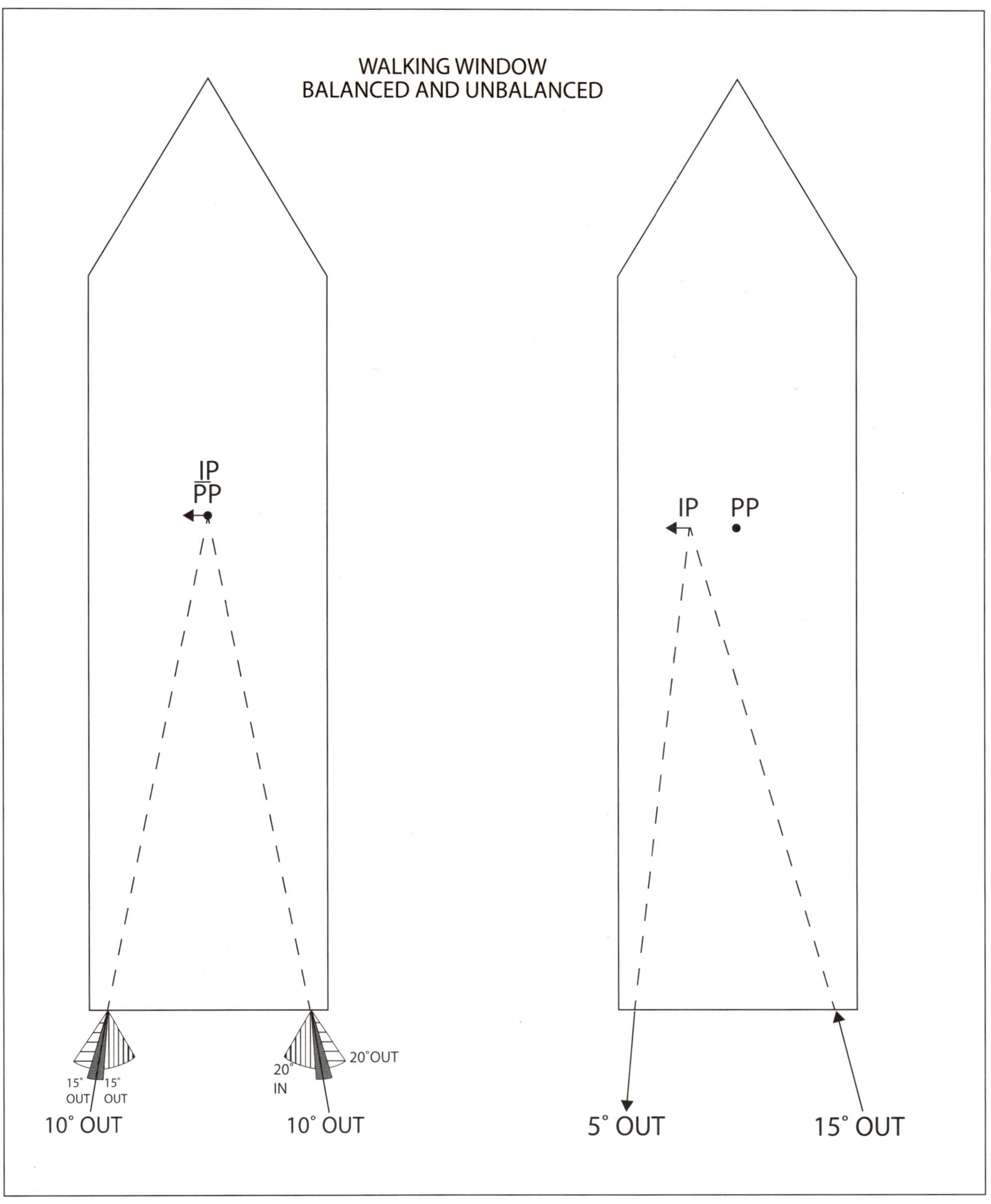

The walking window exists about 5 degrees to either side of the home position. Heading can be controlled while maintaining a walking motion. If either bucket angle leaves the window, the walk will slow, starting a twist. The walking window is the same 5 degrees to either side of home in a balanced or unbalanced walk. Note: in both the balanced walk (*left side*) and unbalanced walk (*right side*), the resultant vector pulls through the pivot point. *Courtesy of Julianne Applegate*

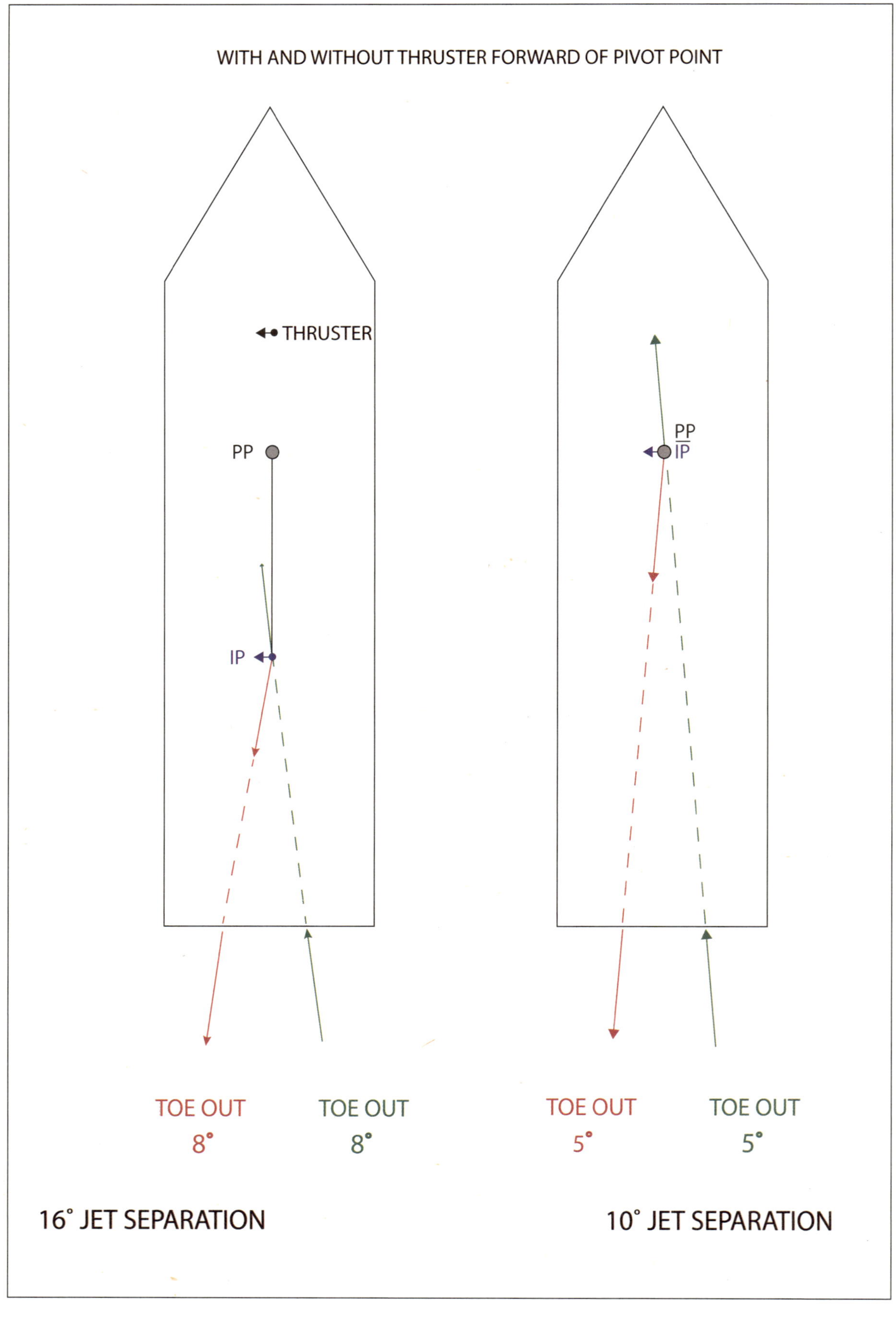

Home position with and without thruster. Using a thruster or tug forward dramatically increases the walking power of a ship. *Courtesy of Julianne Applegate*

Lever arms between the pivot point and the intersection point will induce a twisting motion. Longer arms would mean larger twisting forces for the same resultant vector length. To effectively walk a ship, the intersection point must be kept very close to the pivot point. The angles that keep the intersection point atop the pivot point are known as the home position. Large jet angles toed in or out will always cause large twisting forces. There is a narrow range of angles where the jets are toed out that will keep these two points on top of one another. This range of angle can be referred to as the walking window.

The walking window is best described as the angular separation between the opposed jets. The amount of separation will vary from ship to ship. The easiest way to understand the walking window is to set the toe angles equal to each other, so the angle that the port jet is toed out will be the same as the starboard jet. The toe-out angles of the home position describe the middle of the walking window, with the extended vector lines crossing at the pivot point. As angles are increased (adding toe out), the lever arm will grow, causing the stern to begin to increase lateral speed and eventually becoming a toe-out spin. As angles are eased (reducing toe out, then adding toe in), the stern will slow down, eventually becoming a toe-in spin.

Walking arrangements where the angles are not matched will result in a walk, as long as the total angular separation is the same as the home position. Setting the jets to walk, but not in the home position, will lead to a greater tendency toward ahead or astern bias, depending on which jet is angled closer to the keel. Angles that produce an astern bias will help to lift the bow more easily, accounting for the reduced offset of the jets. Vessels where the distance the jets are offset from the keel is reduced may need to move away from the balanced home position to one where there is an astern bias. For catamarans this is not the most effective way to handle the ship.

The force at the intersection point is a combination of four different sources: the rpm of the ahead and astern jets and the angle of the ahead and astern jets. Trying to control all four of these things at one time is virtually impossible, especially if a thruster is thrown into the mix. To simplify the problem, one set of controls should be fixed, either the rpm or the angles. The primary method of controlling heading is the angle of the jets, so fixing these is not advisable; instead, the rpm should be fixed. The power or speed of the walk is then adjusted by the amount of difference between ahead and astern shaft rpm. Longer opposed vector lengths will produce a higher lateral walking speed. The walk direction is determined by which side of the ship has the astern jet.

WATER JET VECTOR ADDITION (WALKING)

Splitting the plant allows for the astern and ahead vectors, when crossed at the pivot point, to cancel each other out, allowing for a resultant walking vector to be created. Increasing both ahead and astern rpm together causes the length of the ahead and astern vectors to grow. As such, the resultant vector will also grow as either or both the ahead and astern rpm is increased. It is in this way that the power of the walk is controlled. Once the water jet power levels have been balanced for the desired walking force/speed, forward or aft speed is controlled by altering the ahead jet by a small percentage of overall available rpm.

If at any time an engine is lost, the opposing engine will also have to be reduced or stopped to keep things balanced and the speeds in control. Given that the impellers are being driven by high-speed diesels or gas turbines with clutches, it is very easy to quickly depower the impellers to reduce their rpm in the event of a casualty.

Shiphandlers will often be required to bring the ship in to the pier by walking sideways. This means sliding the ship sideways laterally while maintaining the heading of the pier. To do this effectively, the IP (intersection point) must be kept nearly on top of the PP (pivot point). This enables the resultant vector to "push" the ship through the PP, reducing potential twisting.

Once the walk has been created, the speed at which the stern moves is best controlled by adjusting the angles of the jets. The tendency of the ship to move forward or aft along the pier can be corrected by a slight adjustment of the ahead jet rpm in the direction opposite vessel movement.

AHEAD/ASTERN BIAS

Split plant and the slight adjustment of rpm to control forward and aft movement while maintaining a steady walk have been discussed. At some point the shiphandler will need to move forward or astern out of the slip. At this point the balance of engines will need to be upset. To go forward, the ahead rpm should be increased to control forward speed. Ordered speed will not match actual speed, since half the plant is still going astern. Ahead rpm should be limited to the same value as the astern engine; this will ensure that the ahead jet is never near the cavitation limit. Most vessels will be capable of speeds near 5 knots while operating in this fashion. If more speed is required, bring all engines ahead with the correct rpm for the desired speed.

Going astern in split plant presents another problem. When the astern jets are operating near the maximum rpm for cavitation, there can be no increase of astern rpm to generate sternway. The only other option is to reduce the amount of ahead rpm. This will work fairly well until the ahead rpm reaches minimum idle rpm, since at this point the bucket will begin to close (known as pinching the bucket). An immediate reduction in the effective force produced by the ahead jet will be seen. This will result in a reduction of steering control and a quickly increasing amount of sternway. Considering that the pivot point will move toward the stern, reducing the steering arm, using the ahead jet to steer with an already reduced effectiveness can quickly lead to a situation where steering has been effectively lost.

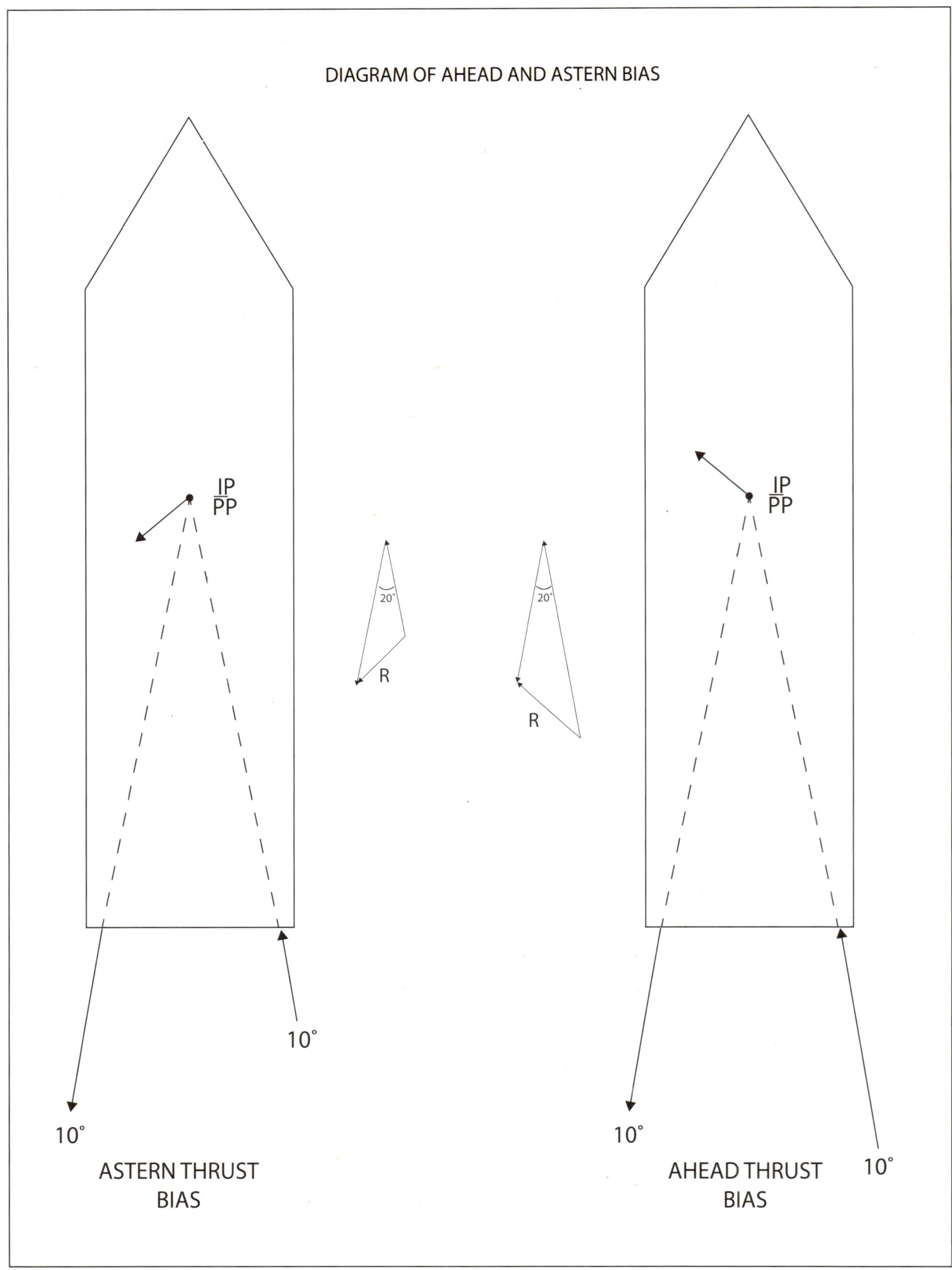

Resultant vectors, showing an ahead or astern bias depending on the power or length of the ahead vector. *Courtesy of Julianne Applegate*

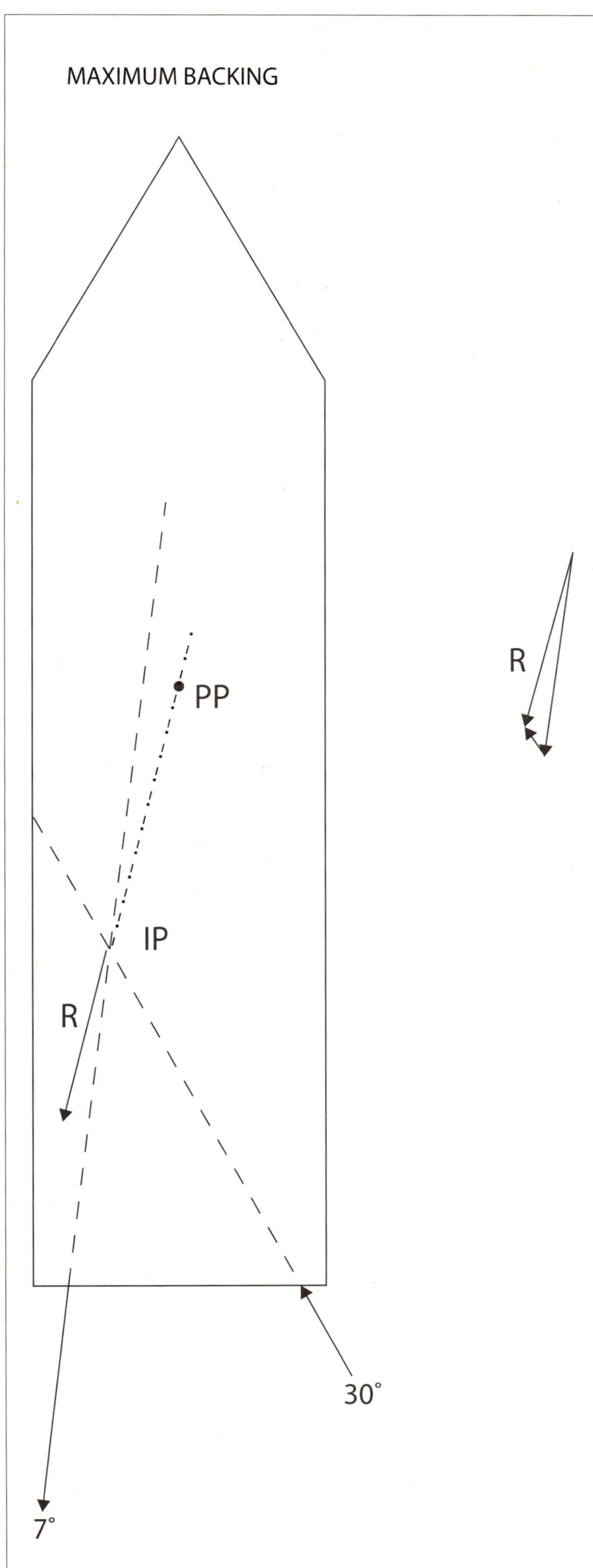

To compensate for the loss of steering control, the ahead bucket should be put to the home position, and steering should be controlled by use of the backing jet. The ahead jet will still be used to control speed by adjusting the position of the backing plate. This will allow the jet with the higher effective wash to be used for steering. Steering with the astern jet is similar to using the ahead jet going forward. If the stern needs to go to port, point the astern jet to port. The stern will always go toward the side that the astern jet is pointed when backing this way, just like the bow always goes toward the side that the ahead jet is pointed when going ahead.

Slight bias, either ahead or astern, is created by changing the angle of the active jet from the home position. As jet angle is toed out, the bias will go in the opposite direction of the active jet. If the angle is toed in, the bias will go with the direction of the active jet. A way to compensate for the change in bias is to increase power when angling out, and to decrease power when easing in. The changes in power required to compensate are very fine and will be most noticeable when controlling the ship by use of the single-jet method.

SPLIT PLANT / BUCKETS

When maneuvering a propeller-driven ship, rpm is directly linked to the speed of the ship. As rpm is increased, speed is increased either in the ahead or astern direction. For a water jet ship this is still true, but there are some differences. When operating in split plant the overall rpm of any shaft should not be equated with vessel speed. The nature of split plant allows for higher rpm on individual shafts, keeping thrust levels and steering forces high but vessel speed at a minimum.

As mentioned, when using split plant the backing jet should be set on the side of the greatest environmental force. Doing so helps keep the IP-to-PP lever arm short, reducing the tendency to twist while crabbing.

Balanced split-plant rpm will prevent ahead or astern movement. When this balance is biased, the vessel will move ahead or astern. The speed is controlled by the level of bias, with higher imbalances leading to higher fore or aft speeds.

When the buckets are opposed but left amidships, there will be a slow toe-in twist. Typically, toeing the astern bucket out to about 5 degrees will correct for this twist and allow the ahead bucket to steer around amidships for going straight.

Split plant provides the captains with the ability to maintain heading control at all speeds. Additionally, splitting the plant once vessel speed is expected to stay below 5 knots, such as in a narrow channel or basin, readies the ship for upcoming maneuvering situations such as docking.

Maximum split-plant backing. *Courtesy of Julianne Applegate*

HOME POSITION

Every ship will have a combinator configuration that will produce an even 0.5-knot walk while maintaining pier heading in an ideal environment. An ideal environment is defined as one where there are no outside forces acting on the ship, such as wind, current, bottom effect, wash reflection, etc. This configuration is known as the home position. The home position will be different for each class of ship. Sister ships may have small differences in the home position due to loading condition or calibration of the jets. All water jet–maneuvering plans should start from the home position.

To properly set the home position, the backing jet should be on the side of the strongest uncontrollable force. Most often the uncontrollable forces that need to be considered are limited to wind and current. (Note: predicted current should not be viewed as accurate enough to determine the backing jet side; rely either on a measurement from a speed log or visual observations.)

The shaft rpm of the astern bucket should be brought to just under the maximum rpm indicated by the cavitation table for ZSTW. At the same time, the ahead shaft rpm should be set to a value that will produce roughly two-thirds of the astern thrust. These rpm values will be very close to those required to hold the ship in position while not accounting for wind or current.

To induce a walk, the jets will need to be toed out. The home position will vary from ship class to ship class and is largely dependent on the distance from the centerline of the ship to the center of the bucket compared to the waterline length of the ship.

The home position is the start point for either the single-jet or matched-angle method of maneuvering. Single jet allows the operator to control the ship as a whole by means of one jet/combinator. The matched-angle method effectively mirrors jet angles across the keel to steer while continuing to control speed with the ahead jet. The matched-angle method really controls the speed and direction of twist and is a more difficult way to handle a water jet ship.

Once the home position has been set, the operator will need to adjust the angle of the ahead jet a few degrees to steer the pier heading and change the power of the ahead thrust up or down slightly to control the position along the pier. Typically the ahead jet will be used as the active jet to steer the walk; the ahead thrust will always be used to control vessel speed, since the astern rpm is at or near its maximum value as found on the cavitation table.

Knowing the home position for the ship being driven is most important for when the operator loses awareness of the maneuvering situation. In the case where nothing is working as it should, return to the home position to rebaseline the maneuvering inputs and settle out the ship. This requires a good deal of patience and confidence. Additionally, sufficient time and distance must exist from potential dangers.

CONTROLLING THE CLOSURE SPEED ON THE BOW AND STERN

When using the home position to walk, the bow may come away from the initial heading. This can primarily be seen by observing relative motion and bearing change visually. As a backup, some ships are fit with rate-of-turn indicators or speed logs. These tools are extremely helpful, but it is incumbent on the shiphandler not to overly rely on them, since they tend to lag the actual motion of the ship by a few seconds. If the shiphandler tries to make inputs on the basis of the indications as opposed to the observed motion in the window, they will always be behind the ship and chasing their tail, resulting in overcorrections.

The best way to avoid either end of the ship from closing the pier too quickly is to keep the ship on pier heading with a minimal rate of turn—less than 2 degrees per minute. This will ensure that the ship lands flat along the pier.

If the stern gets going laterally too fast, there are two ways to slow it down. Shift the ahead bucket until the stern-walking speed begins to slow, then go back to an angle slightly closer to home than before the shift. The other option is to shift both buckets evenly. This will rapidly slow, stop, and potentially reverse the stern-walking speed. Using both buckets in this manner, if not done with dexterity, will quickly lead to progressive overcorrection.

If the bow gets going too fast laterally, the solution is very similar to correcting the stern speed. Shift the bucket, then ease back once speed is under control. Shifting both buckets is also an option in this case, with the same potential effects as described with slowing the stern-walking speed.

The amount of the initial shift is determined by how fast the ends of the ship are moving: the faster they are going, the more angle that will be required to arrest the speed in the same amount of time.

In every case the walking speed of the bow, stern, and ship in general should be at or below the maximum landing speed with the vessel on pier heading when less than one ship width from the pier.

BOW THRUSTERS

Typically only larger ships are equipped with bow thrusters. Ships with bow thrusters will be able to handle increased environmental forces compared to a similar-sized ship lacking thrusters.

Just as there is a lever arm between the IP and PP, there is also a lever arm between the PP and the bow thruster (BT). Unlike the lever arm for IP, the arm between the thruster and the PP cannot change in length unless the ship's overall headway or sternway is changed. The addition of a second lever arm acting on the PP is responsible for the increase in maneuverability.

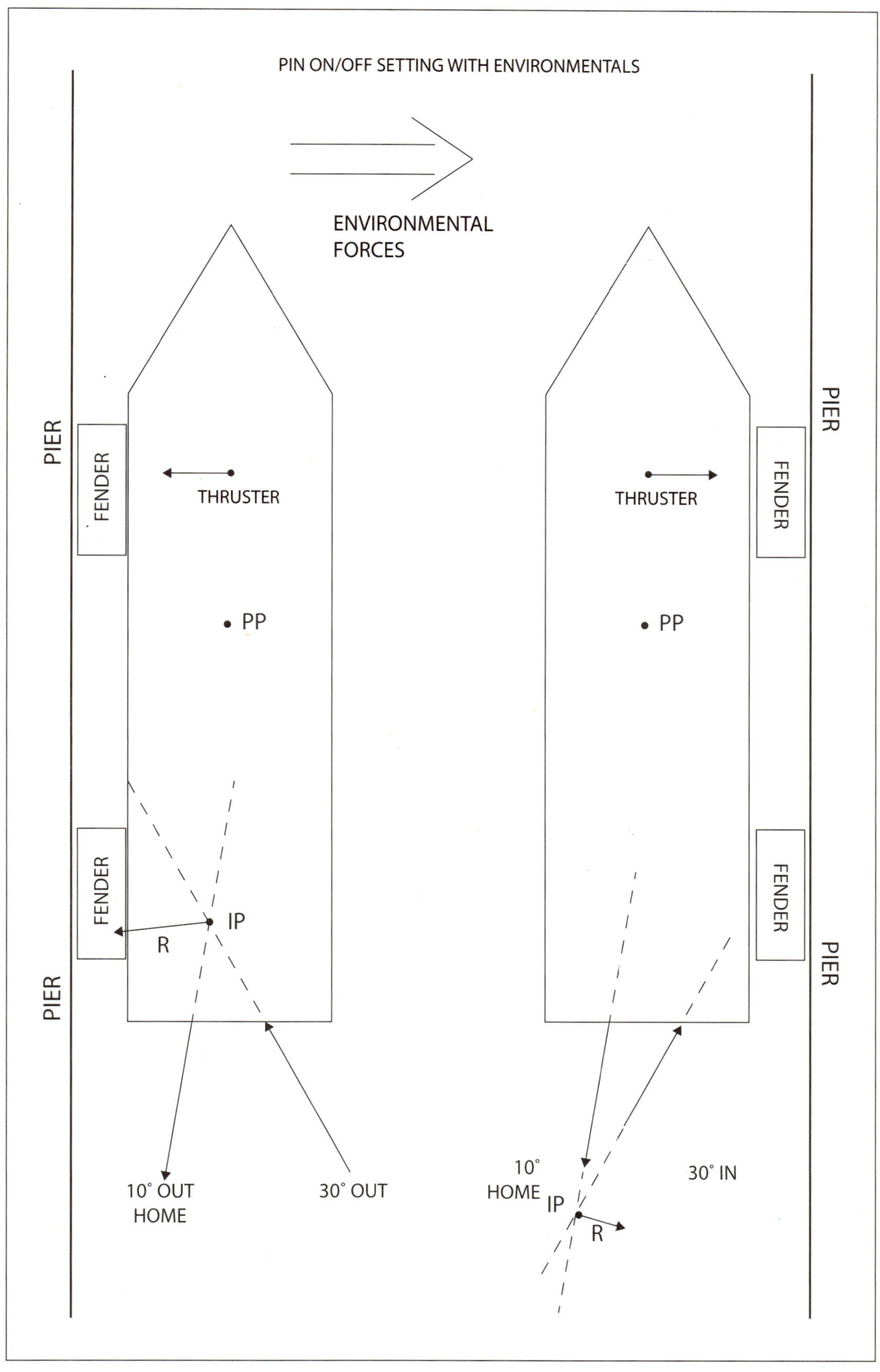

Pinning with a thruster both against and with environmental forces. The thruster or tug will always push the bow toward the pier. This allows the backing jet to be on the side of the strongest environmental forces. The ahead jet is then toed to drive the stern into the pier. Wind in (toward the pier), toe in / wind out (away from the pier), toe out. Note: if the current is stronger, replace wind with current and apply the same concept. *Courtesy of Julianne Applegate*

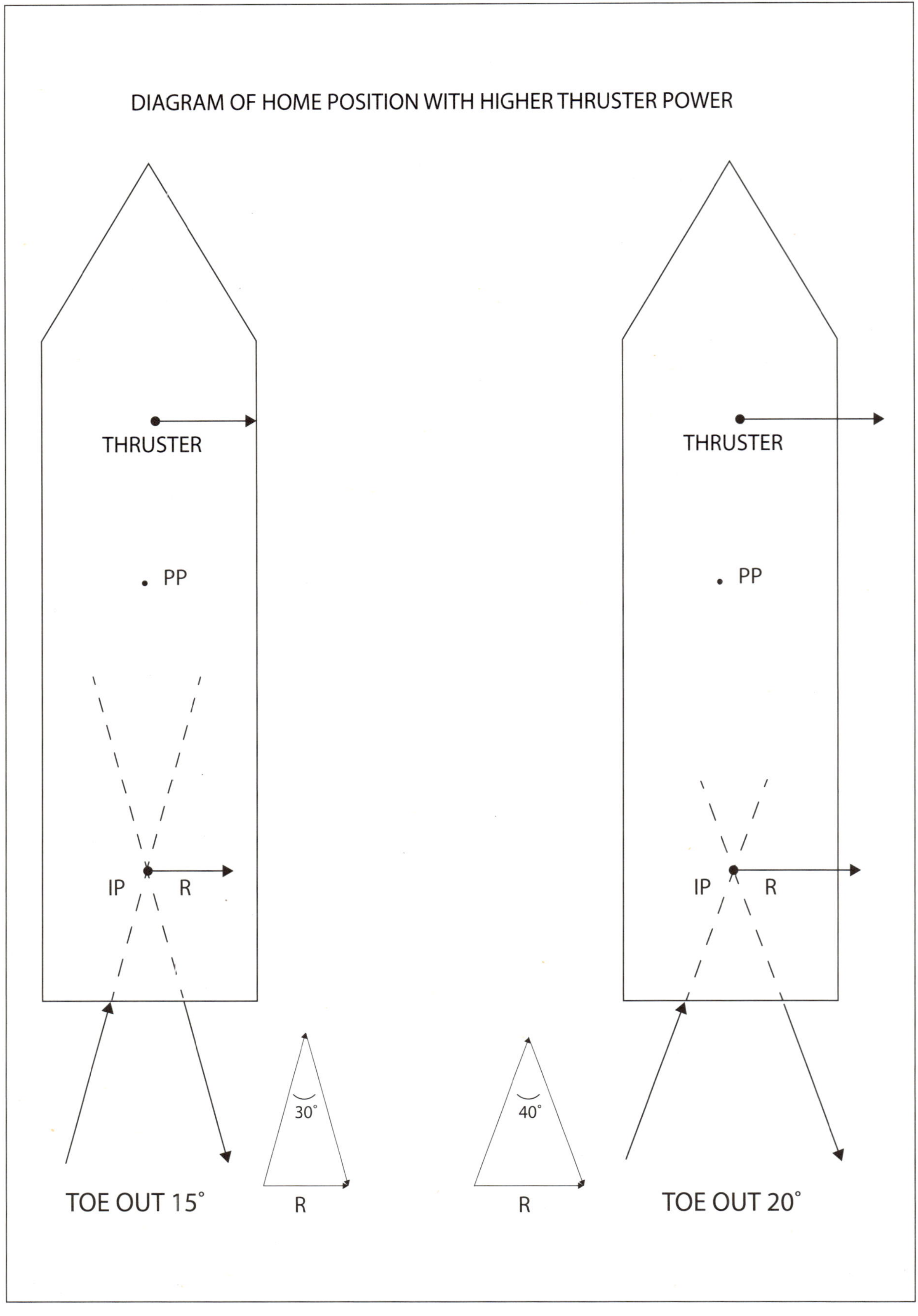

Increasing the power of the forward thruster or tug will require moving the intersection point aft to maintain a balanced walk. Further toeing out both jets will achieve this. Normally only the active jet will need to be toed out to maintain heading, and a balanced walk. *Courtesy of Julianne Applegate*

The length of the BT-to-PP lever arm is effectively fixed as long as the ship is maintaining a constant speed. To balance the walk, the lever arm between the IP and the PP must have equal force. This does not mean they will have equal length. Think of a seesaw or balance where one side is a fixed length but has a weight that can be adjusted, while the other side has a constant weight that can be moved along the arm. Setting one side of the balance and then adjusting the other side to even things out as environmental forces act on the balance would be very close to how the ship will now be controlled. Setting the power of the thruster and then moving the IP second is the easiest way to find the balance point of an even walk.

There is still a home position when using a bow thruster; the difference is that use of the thruster allows for the angles of the jets to be toed out more from the keel to shift the IP aft, away from the PP. If this shift does not happen, the ship will walk while the thruster spins the ship. Setting the home position starts with the bow thruster. The bow thruster controls the speed and direction of the walk at all times and is adjusted to counter the observed environmental forces. Once the thruster has been set, the jets are taken to the home position. While walking, heading control is still a function of the ahead jet angle, and speed is still a function of ahead jet rpm.

While this may seem like more individual forces to keep straight than a ship without a thruster, it is in fact easier to control a ship with a thruster once the process is clearly understood.

PINNING

Traditional ships will make use of tugs to lean, or pin, a ship into the pier as the lines are let go to keep the ship under control. Water jet ships will not typically use tugs and, as such, will need to find another way to pin to the pier.

Pining to the pier has two primary benefits. First, the ship's crew are able to set engine rpm for the conditions prior to letting go. Second, a ship leaning on the pier will not tend to move as much, making line handling safer for the deck crew.

Pinning without a thruster is a matter of splitting the plant in home position, with the backing jet on the pier side of the vessel, then angling the ahead jet outside the home position to push the transom into the pier. Until a vessel is well understood, power should be brought to that of a maximum walk. Once a vessel is well understood, power should be set to that needed to walk into the prevailing offsetting conditions, or home power settings for onsetting conditions.

Pinning with a bow thruster is different than without. Vessels with a thruster will set the engines to max-power home position with the backing jet to the environmental forces. The thruster is directed to push the bow toward the pier. The ahead jet is then angled to push the stern into the pier. For offsetting conditions the jet is toed out; for onsetting conditions the jet is toed in. If offsetting can be seen as "out"-setting and onsetting seen as "in"-setting, the following applies: outsetting toe out to pin, and insetting toe in to pin.

To unpin either type of ship, walking power toward the pier is reduced to see if conditions will lift the ship from the pier. If power has been reduced to a minimum level and no lateral motion is observed, the direction of the walk is set away from the pier.

Ships without a thruster will reduce the amount of split between the ahead and astern rpm until lateral motion away from the pier is observed while steering pier heading. If power is reduced to a minimum level with no motion away from the pier, the split is reversed and power is increased slowly to walk the vessel away from the pier.

Ships with a thruster will reduce the power of the thruster slowly until ROT (rate of turn) away from the pier is observed. At this point, use the active jet to steer pier heading to lift the stern. If the thruster is reduced to a minimum level, direct thrust away from the pier and increase slowly until ROT away from the pier is seen, and use the active jet to lift the stern. Waiting for ROT ensures that the thruster is correctly adjusted so that the ship will walk away from the pier. Most vessels can lift their sterns away from the pier in all but the most extreme conditions, and as such, if the thruster is not properly adjusted, the bow may not lift, complicating the docking maneuver.

GENERAL RULES FOR HANDLING ALONGSIDE

Be mindful of speed. Sometimes speed is a plus, while other times it is a negative. Know the difference.

Plan ahead and constantly consider if the ship is going where desired.

Getting away from the pier should always be an orderly process. Ease power toward the pier before setting power away from the pier. This will prevent the ship from being sent flying across the slip.

Due to the relative lightness of water jet ships, it is better to positively push the ship where you want to go rather than allowing the ship to glide to a stop. Remember that while operating in split plant, half the jets are trying to stop the ship.

When coming out of a transition, push the ship in the general direction you want to go, and do the fine tuning once you have built up some momentum.

Always put the ship in the middle of good water, which may not always be the middle of the slip.

Provided there is sufficient room, conduct twists as fast as possible. Less time will be spent at the mercy of environmental forces, and less time will be spent blocking the channel.

Twisting the bow into the wind or current is always preferable, even if the stern will be closer to danger than the bow. If the vessel is set down on an obstruction, ahead power can be applied to move away. If the bow is close to danger, ahead power has to be eased, resulting in reduced effective wash through the ahead jet.

The heading of the pier is critical to understanding the position of the ship in the slip and alongside the pier. Staying parallel to the pier is always a good idea.

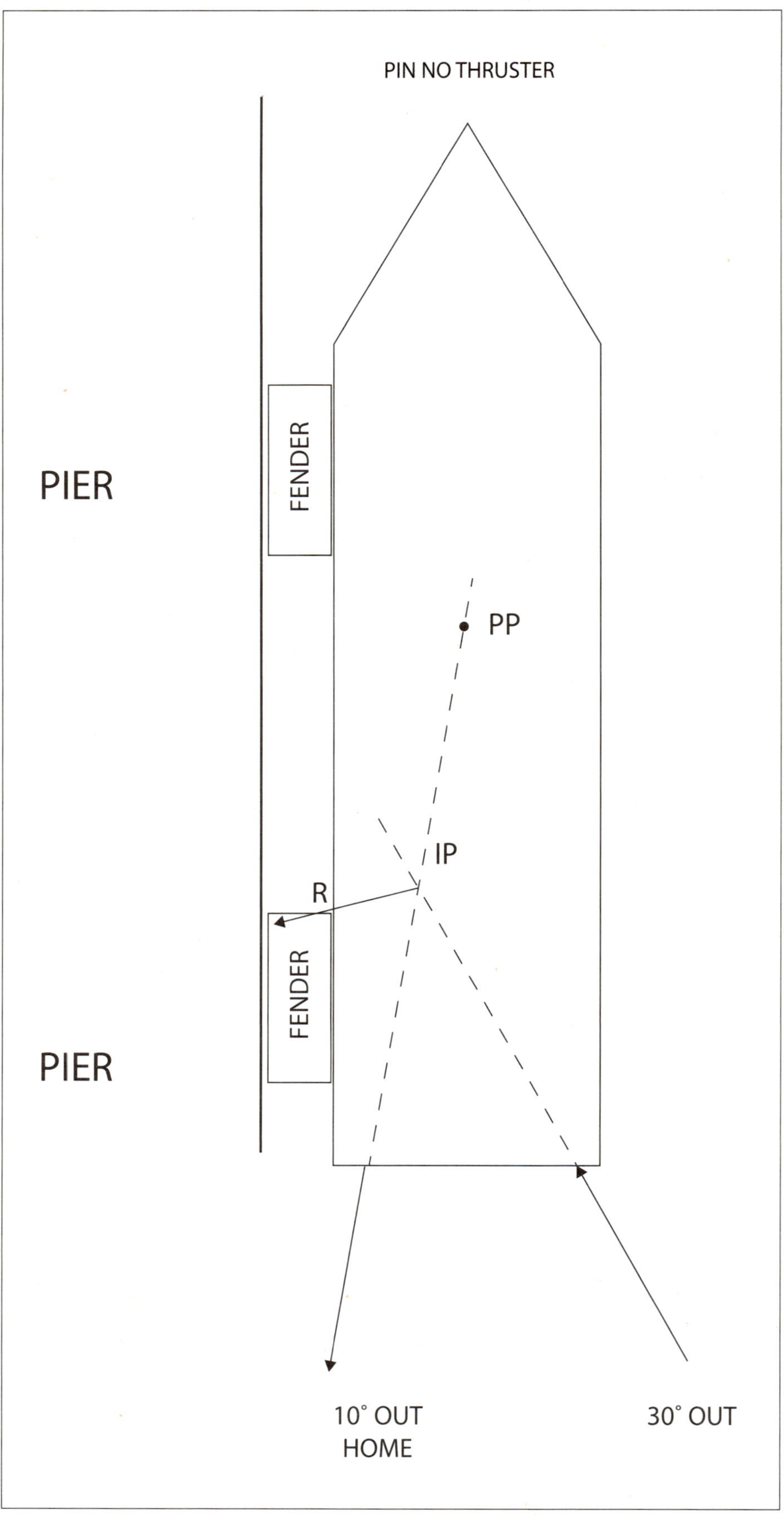

Pinning to the pier without a thruster. The backing jet must always be on the pier side of the ship. *Courtesy of Julianne Applegate*

SINGLE HULL VERSUS MULTIHULLS

There is considerable difference in the things that contrast single hulls and multihulls: stability concerns, fuel consumption, maximum cargo weight, and interior volume are a few examples. The main issue to be discussed is how they handle alongside the pier.

Catamaran jets are pushed out to the extreme breadth of the ship. With additional width verses length, twin hulls are able to use more toe-out angle. This allows catamaran vessels to walk against higher environmental forces.

Trimarans do have multiple hulls, but the fact that the jets remain in the center hull means that the toe-out angles are similar to a monohull. In reality the center hull of a trimaran is typically very narrow, further reducing the steering level arms and virtually requiring a thruster forward to do any shiphandling in increased environmental forces.

The addition of a thruster completely changes how shiphandling is approached. The relationship of the pivot point and the intersection point remains a critical point to understand. However, the lever arm is not kept as small. The thruster position along the hull produces a lever arm. This lever arm must be balanced by the lever arm from the intersection point. The resultant vector no longer going through the pivot point for a walk changes the relationship of walking speed to vector length.

The largest change in ship handling is understanding that the speed and direction of the walk is no longer defined by the power and direction of the resultant vector; it is now defined by the power and direction of the thruster. This sets up the strength of the lever arm acting on one side of the pivot point. The lever arm from the intersection point must have an equal force for the ship to walk evenly.

To continue to keep things simple, rpm should be set so that max power is available at any time only when using a thruster or tug forward. The astern jet should be set at or just below the ZSTW cavitation point. This allows the ahead bucket to be opened, maximizing the flow and steering of the ahead jet. This fixes the length of the resultant vector.

The length of the lever arm is then controlled by the angle of the jets. To further simplify things, only the ahead jet is used; the angle of the astern jet is fixed. The astern jet is referred to as the static jet, and the ahead jet is the active jet. The length of the intersection point lever arm is controlled by the angle of the ahead jet.

Operating the engines near 100 percent of available rpm also provides that all power is at the ready in case of an emergency. An increase in the level of wash exiting the jets is a concern. Having the astern jet operating near maximum rpm allowed by cavitation will cause a significant increase in wash over that which presents when at idle rpm. Since this wash is directed under the ship, it should not present problems to other vessels in the harbor. The wash will have an impact on the bottom, and increasing rpm will cause more bottom to be stirred up. The whole reason to have the astern jet at or near maximum rpm for cavitation is to be able to fully open the ahead jet, since ahead rpm will be only slightly over idle. Therefore there will be no appreciable increase in the amount of wash coming from the stern of the ship. However, there is still quite a bit of wash exiting the stern, requiring care on the part of the operator to limit the effects on other vessels and port structures.

CHAPTER FOUR

HIGH-SPEED NAVIGATION AND BRM

BRM (bridge resource management) for high-speed bridge teams functions differently than on traditional ships. On traditional ships, information is designed to flow to the person in charge of the bridge—the captain or watch officer (for simplicity, the person in charge will be referred to as the captain). Orders or requests for information flow from that person. Mainly the captain is trying to use all available resources to solve the navigation picture on their own.

A comparison can be drawn by thinking about a car. Traditional BRM has the helmsman driving the ship/car. The lookout sits in the front passenger seat, looking for danger. The navigator is behind the helmsman in the back row, occasionally looking through the window to double-check the position. Finally, the captain is behind the lookout, keeping an eye on the helmsman while looking out the window to verify that the ship remains out of danger. A well-trained and practiced team would have little problem navigating their car from one side of a parking lot to the other. As both parked and moving cars are added to this problem, the ability of the team to function quickly enough to control the car rapidly degrades. There are reasons we do not drive cars this way, just as there are reasons of ship design that have required ships to be driven this way.

Modern automation has allowed for direct control not only of the helm, but also the engineering plant at the conning position. Radars no longer require appliance-sized cabinets with space for cooling around them, allowing for these units to be positioned at the conning station. The advent, and requirement, of ECDIS use aboard ships has reduced or eliminated paper charts. Since large chart tables are no longer needed, the vessel's position can be directly monitored from the conning station instead of a separate chart room. The integrated bridge system allows for a complete change in bridge layout and BRM utilization.

High-speed BRM endeavors to move the captain into the driver's chair while keeping the other bridge team members in the car, where they can be effective resources. Much like a car, this allows the team to function quickly enough to navigate the vessel while still maintaining high levels of safety and situational awareness.

Higher transit speeds naturally suggest that there will be increased amounts of information to be processed by the watch team. Traditional styles of BRM and communication do not work well in these situations. Typically, larger high-speed ships will be designed so that there is a two-person navigation team. The bridge design may also locate an engineering watch station on the bridge. These three team members make up the bridge team.

When all three team members work together, information flows in a more conversational way. Conversing about the navigation picture shares each team member's own mental model. Shared mental models do a few things for the team. The desired end goal is clearly stated. The information being used to form the mental model is discussed quickly, catching errors or omissions in situational awareness. When new information that changes the mental model is presented, all team members can work together to quickly review the situation and find a solution. Each team member has a set of responsibilities that at times will overlap. It is up to the captain to responsibly manage available resources and requirements to meet the needs of the current situation.

Many high-speed vessels are catamarans, and the engine rooms on these vessels are large enough only for the engines themselves and their associated equipment. There is no room for an engine control room. Even if there was, the two rooms are typically separated by a cargo deck, with no place to efficiently design in an engine control room. With modern engine management sensors and controls, it is now possible to station the on-watch engineer on the bridge. Having the engineer form part of the navigation watch offers immediate information to the captain when dealing with normal operating conditions and in the event of a mechanical failure. The engineer is also able to take control of the various fire suppression systems and dewatering pumps, providing better response and control in the event of a vessel emergency. Locating the engineer on the bridge does present a few problems for effective engine room management. These are offset by using radios and the positive benefits to the overall operation of the vessel.

The captain's assistant is best referred to as the navigator. The navigator is responsible for double-checking the current and future position of the vessel, maintaining lookout by sight and radar—often using the radio to communicate with other vessels—and any other action that would assist the captain. They are most concerned with where the ship will be in the next twelve to eighteen and possibly twenty-four minutes.

The captain is always responsible for the safe navigation of the ship and is the final say on how the navigation picture will be solved. As the lead team member, they are responsible for overseeing the other members of the team, keeping track of the vessel's current and future position, operating the controls for speed and heading, and maintaining an effective lookout both by sight and radar. They are most concerned with where the ship will be in the next six to twelve minutes—the smaller, more immediate picture.

While both the captain and the navigator are maintaining the lookout, they are doing it in different ways. Both will make use of radar, and both will look out the window. The ratio of eyes out the window is expected to be different for each. The captain should be looking out the window 75 percent of the time. At no time should the captain be looking down for more than about ten seconds.

At 30 knots a vessel travels 3,040 ft./minute; for larger vessels this could be 10 ship lengths. This is close to the range that partially submerged shipping containers, small fishing vessels, fishing nets, and large logs are typically detected at, often by eye and not radar. Assuming the watch stander saw any of these at a mile, a good chunk of the mile is used just trying to figure out if that thing in the water is a danger. Once a decision is made, the reaction time of the ship further reduces the final CPA (closest point of approach) to the dangerous object. If the time to CPA is further reduced by a poor visual lookout, the final result may not be a near miss, but a collision. Since the captain has their hands on the controls, or very near them, they are the best and last person to take evasive action in the event of an extremis situation.

The navigator should not being looking up as frequently as the captain, typically around 50 percent of the time. They are more correctly concerned with the larger picture. Traffic out ahead, possibly beyond visible range, often shows up as an AIS contact and not a radar return. An effective navigator feeds the information on new contacts to the captain so that contacts do not enter the smaller area of concern without notification. In a sense, there should be no such thing as a pop-up vessel.

EYES UP/DOWN

It will not always be possible or advisable for the captain to strictly maintain the ten-second rule for looking down. When there is a need for the captain to look down for more than ten seconds, the navigator will need to be instructed to take over visual lookout duties. This is accomplished by the captain stating, "Eyes down." Once the navigator has acknowledged with "Eyes up" and is actually looking out the window, the captain is free to look down to accomplish the needed task. Once the captain has completed the task requiring eyes down, the navigator is informed that responsibility for the visual lookout can revert back to the captain with "Eyes up." Using this terminology allows the bridge team to efficiently and effectively maintain a proper visual lookout at all times.

NAVIGATION AT SPEED

The modern bridge makes use of large screens to present radar, ECDIS, and other information. The high-speed bridge often goes a step further, removing the chart table and paper charts. The reasons are simple. Assuming that a navigator could record and plot a fix on a chart in thirty seconds, at 30 knots the fix is already 1,500 ft. behind the ship. This is more than a ship length for all but the largest ships. An additional reason is that the actual movement of the vessel often makes accurate chart work impossible, since there is too much vibration and motion.

The increased amounts of motion and their causes have been mentioned. As sea state increases, the ability of the watch stander to actually "stand" the watch lessens. So much time is spent just trying to remain standing that little time is spent navigating. Bridge chairs are part of an integrated bridge system design for this reason. Enabling the watch to sit rather than stand allows the team to focus more attention on navigating. Unfortunately, this design feature makes it hard to locate an effective visual-bearing repeater near the chair. Often these repeaters are away from the conning station, where they are of limited use. Effective use of EBLs on radar can mitigate this problem. Requiring watch standers to sit in one place has the benefit of making other ships' relative motion easier to track against the window or console or between window stanchions.

The radar installed on high-speed ships may be standard units, or they may be high-speed radar. The difference is the antennae rotation rate. High-speed units rotate at twice the speed of regular units. While this reduces the maximum effective range of the radar, the average mounting height on most high-speed ships prevents the use of maximum range anyway. The increased rotation rate leads to quicker target acquisition and solution for relative and true vectors. Instead of the fairly common three to five minutes for a solution, these radars solve the vector in one to three minutes, reducing the effects of ARPA lag.

The advent of ECDIS and its use on high-speed ships has changed how the navigation problem is approached. Continuous updated positioning allows the watch stander to view the chart in near real time. This allows for more-accurate DRs and better understanding of set and drift. When radar or AIS target information is overlaid on the chart, a bird's-eye view of how traffic is affecting the problem can be seen in as close to real time as possible.

It is very easy to slide into the mindset that the screens present a better version of reality than what can be seen out the window. What is happening outside the window is always the current situation and is accurate and changing by the second. The information presented on the individual screens must be gathered, processed, distributed, and presented, to then be observed by the bridge team. That whole process can take seconds for each individual bit of information. However, some of the information presented is partially based on information obtained from other sensors, adding more delay. Timeliness of information is just as important as the accuracy and must be considered in assessing the value of the information.

The amount of information that can be processed at one time is referred to as cognitive burden. The level of maximum burden is different for each watch stander, depending on age, experience, ability, training, and knowledge. Cognitive burden can be equated to stress. Healthful levels of stress keep watch standers involved and interested in the current situation. When the maximum—either stress or cognitive burden—is reached, the brain naturally stops trying to deal with the bits of information that are deemed extra, and situational awareness decreases.

For many, the first thing dropped is maintaining a proper lookout, preferring instead to seek answers in the information screens, believing that more-correct and up-to-date information is presented by the computer. One term for this is the "magic-box syndrome." Others will focus solely on the window, paying no attention to on-screen information.

At slower speeds this is not much of a problem, since time and space usually allow bridge teams the gap needed to regain situational awareness and work their way out of the problem by using only one set of information. The quickness at which the distance to danger is reduced at high speed requires that bridge teams make use of all information while trying to solve the navigation situation. The reaction to this close danger will have implications on situations that will happen after clearing the first danger.

MAXIMUM SAFE SPEED

If situational awareness cannot be quickly regained, it is necessary to slow down to reduce the speed and the amount of information that the bridge team is being presented with. This can help define the difference between traveling at maximum speed versus maximum safe speed. Maximum speed simply means going as fast as conditions allow, with no regard to safe navigation. Maximum safe speed is the highest speed at which the bridge crew can take proper and effective action to avoid collision and be stopped within the distance of prevailing circumstances and conditions.

Clearly there is more to maximum safe speed than simply current visibility compared to the vessel's stopping distance. Other circumstances considered should include traffic density, underkeel clearance, bottom profile, distance to shore, type of facilities present on the shoreline, visibility and stopping distance, and, finally, the bridge team's ability to keep making proactive decisions in the given situation.

Indications the ship is proceeding above maximum safe speed:

- There is loss of situational awareness.
- Decisions become increasingly reactive.
- CPA distances are below the minimum safe threshold.
- Large wakes are being created.
- The situation devolves to the point where full astern is used to slow to avoid a collision.

This list does not represent the only indications that a vessel or the bridge crew is going too fast, but they tend to be the most obvious ones.

Figuring out just what maximum safe speed is also requires looking at what the indications of going too slow are. The point of high-speed vessels is to get to the other side, quickly. In this vein, if the bridge team has full situational awareness but low levels of chronic unease, then maximum cognitive burden has not been approached and the team has the ability to increase speed. Another indication of going too slow is not gaining on traffic ahead. Other ships typically cannot achieve the same speeds as water jet vessels, and as such, if they are not being overtaken, the team should evaluate why the ship is going so slow.

Reducing power when a new issue is detected, in order to think, then almost instantly bringing the power back up, is an indication of poor situational awareness. This is called thinking with the throttle and is not the same as losing awareness. In this case the various systems are not being properly used to get ahead of the problem. The issue here is not speed, but an individual operator's own abilities. If the speed was okay before the stimulus and then okay after a few seconds of thinking, then there was no need to slow down.

When a ship is operating at maximum safe speed, the following should be evident:

- Proactive choices are being made.
- There is full awareness of the navigation and traffic situation.
- The team can solve the problem without slowing down to think.
- There is a healthful level of chronic unease.
- The mental model is fully shared.
- CPAs are maximized above minimum thresholds.
- Distance from land or shoal water is maximized.

Keeping a ship at maximum safe speed is a lot like walking a tightrope. Too fast is just that, and the ship needs to be slowed. Too slow will not get the ship to the next stop on time, and that is why a water jet ship is being used.

ADVANTAGES AND DISADVANTAGES OF HIGH SPEED

Traveling faster than other ships in the area can have its advantages. First, speed can be adjusted to control the time and distance of various CPAs. This means that the captain of a faster vessel is in control of most meeting situations and is not as reliant on other vessels. Second, traveling at higher speeds means less overall time is lost giving way to other vessels. Due to closure rates and the poor maneuvering ability of other ships, it is often more beneficial for the high-speed vessel to give way early, even if they are the stand-on vessel. However, giving up the right-of-way must always be done with proper caution and good communication between vessels.

While there is nothing in the rules of the road that refers to the speed difference between meeting vessels, being in control and making decisions early can change how the rules are applied. If action is taken well before risk of collision exists, then almost any action is permissible. Frequently a situation will present itself where an alteration of 5 degrees to port and increasing speed will result in an equal or greater CPA, compared to maintaining speed and altering course 70 degrees to starboard. In these situations it is critical for the watch team to remember that once action to port is taken, the heading cannot be allowed to come back to starboard until past and clear of the other vessel. If further action is required, slowing down or continuing to port are the only options available; anything else will be too confusing.

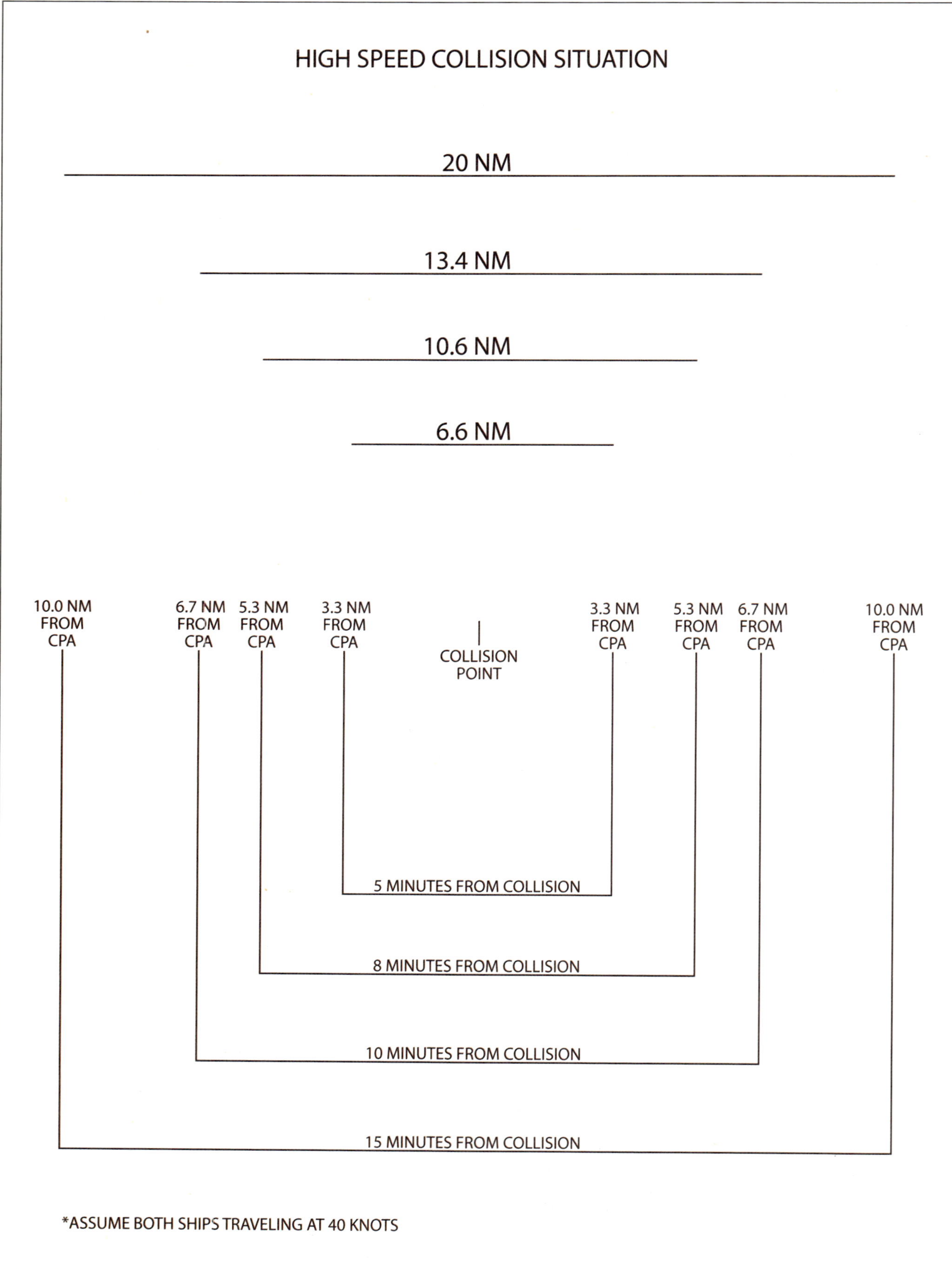

A comparison of TCPA to range, showing how little time is available to process a potential risk of collision when two high-speed vessels are approaching one another. *Courtesy of Julianne Applegate*

The situational awareness of high-speed bridge teams is almost a half hour into the future. While slower ships may have similar awareness in terms of time, the distances being looked ahead are often not the same as for faster ships. This would mean that high-speed ships are often aware and ready to take action earlier and at greater distances than regular ships. Action taken in this way happens long before other vessels have even ascertained that risk of collision ever existed, and eliminates the need and ability of other vessels to maneuver, even when they are to give way.

As with anything, there are advantages and disadvantages to traveling fast. The biggest issue is that closure rates between vessels are much higher, reducing the time to makes decisions. Making things worse is that often the radar antennae heights on high-speed vessels are not as high as on other ships, reducing the radar horizon. Most designs also reduce the height of eye on the bridge, thereby reducing the visible horizon distance. The reduced horizons affect look-ahead and the available time for making decisions. Two high-speed ships closing on one another further compounds this issue.

Working with an assumed radar horizon of 20 nautical miles (NM) for two ships traveling at 40 knots, an 80-knot closure rate would mean that each bridge team has fifteen minutes to avoid a collision. If both ships acquired the other 20 NM apart, the ARPA would then take about three minutes to plot a vector. The team would need an additional two minutes to consider the situation and possible solutions to avoid the other vessel.

At this point the two ships are ten minutes from collision but 13.4 NM from each other, and the two bridge teams still need to communicate with each other. The easiest way would be to simply alter course to the right enough to get a 2 NM CPA. If instead the radio is used to confirm intentions prior to altering course, an additional two-minute delay would be added before a maneuver took place. The ARPA plot would not display the new situation, after the maneuver, until five minutes before collision, when the two ships have closed to a separation of 6.6 NM. In this example the total time elapsed from initial detection to observed avoiding action is ten minutes.

At these speeds, for every minute after the ships could have spotted each other at 20 NM separation that elapses before the avoidance process starts, the distance between the ships is reduced by 1.3 NM. It is more likely that the two ships would "see" each other about 18 NM apart. This leaves about fourteen minutes to get through the avoidance process, which takes around ten minutes. On the basis of the avoidance process taking about ten minutes, the bare minimum distance apart these ships can be and have enough time to avoid a collision in an orderly fashion would be 13.4 NM, and that gives the bridge teams only one chance to get it right. This amount of time falls just about between the navigator's and captain's responsibilities in terms of looking ahead. Once you're inside ten minutes, any avoiding action becomes more of a reactive choice rather than proactive and can quickly become an extremis situation.

ERROR CHAIN VERSUS CHAIN OF SUCCESS

Reactive choices result from having limited time, distance, and information to make fully informed decisions. A team making a series of these type of choices is said to be behind the problem. They are not keeping up with the rate of information being presented to them. They are going too fast for the situation or themselves. Failing to recognize the types of choices being made and reducing speed will be viewed as one of the larger links in the error chain in the event of an accident.

Being able to plan ahead means that there is enough time, distance, and information to make an informed decision, requiring full situational awareness of the world around the ship. Any loss in awareness can shift choices from proactive to reactive. Loss in awareness can happen from continuing to go too fast in a changing situation, a reduction in the distance to danger, or the loss of a source of information. Choices made by planning ahead are seen as proactive choices, which are not links in the error chain and tend to lead to a success chain.

Error chains are ways of linking various events and choices to understand how an accident happened. Each link in the error chain represents a point at which the chain could have been broken. Breaking the chain at one point would not have prevented the accident, but it would have reduced the chances that the accident would have happened. Break enough links and the odds are fairly low that an accident happens at all. Success chains aim to build on that logic. If enough decisions are made that reduce the possibility of an accident, the odds of success greatly increase. The simplest way to describe success chain choices are in terms of risk and reward. Success chain choices seek to reduce risk and increase reward, setting up a decision process that prioritizes low-risk, high-reward choices. For vessel operations this means always positioning the ship so that risk is minimized to the largest extent possible.

Following are a few examples of setting up for success:

- Have the various information systems on the bridge set up so that they provide clear, concise information in a way that they can be useful while only being glanced at, and require minimal manipulation by the watch stander.
- Steer for the middle of the good water at all times.
- Keep as much of the danger (ships, shoal, shore) on one side of the ship as possible, which will reduce the workload on the bridge team.
- Stay ahead of the problem.
- Make choices that result in high reward with low risk, not low reward with high risk.

These steps reduce cognitive burden and allow the team to function more efficiently with less stress, thus reducing the effects of complacency by practicing a healthy amount of chronic unease.

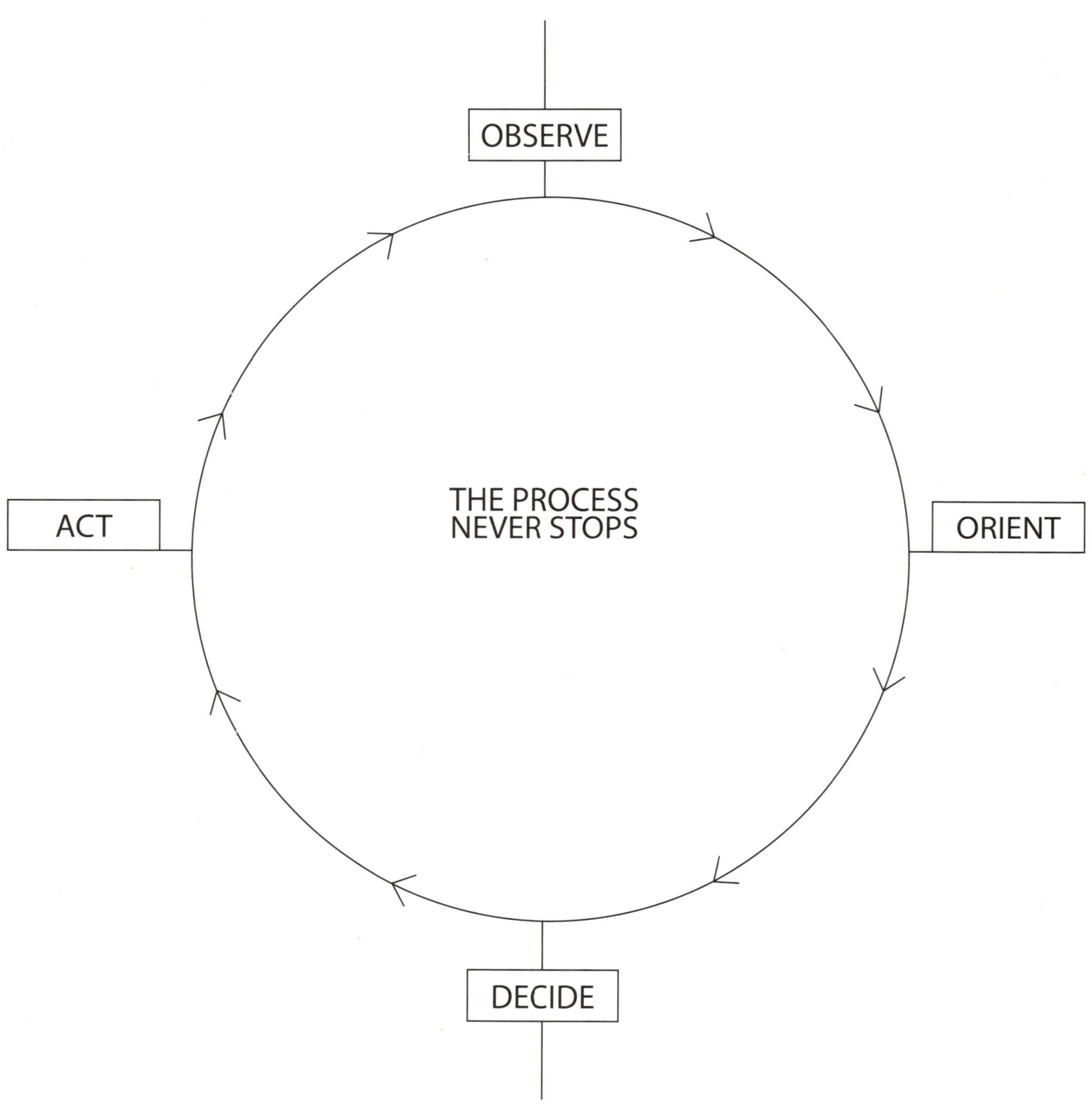

The OODA loop is used to continually work toward improving situational awareness, with the information available as quickly as possible. *Courtesy of Julianne Applegate*

Complacency happens when stress levels are allowed to fall below the minimal level needed to remain interested in the watch. There are some common signs of complacency: not feeling the pressure of the task at hand, assuming that the task is under good control without looking for potential problems, feeling free to focus on nonwatch tasks, excessive nonwatch conversation with other members of the bridge team, and a feeling of overconfidence. The scariest thing someone can say in regard to complacency is "I'm really good at my job and have things totally under control"[1]

The exact opposite of complacency is chronic unease. On the surface, it may seem that this is just as bad as complacency. Chronic unease is the term used to describe a healthy level of wariness. A general feeling of concern that something is being missed, and therefore not all the information needed to solve a problem is present. Clearly, if this idea is taken too far, to unhealthy levels, it would be debilitating. In real life all the information needed to solve problems is never present. This doesn't prevent decisions from being made. Allowances are made to offset this, at times a small lack of information. Healthy levels of chronic unease can also be referred to as a questioning attitude. When additional information is not always being looked for, the one piece of information needed to change the mental model and prevent an accident will not be observed.

The US Air Force teaches pilots to continually observe, orient, decide, and act (OODA). Done over and over as a cycle, this is called OODA looping. This process, coupled with chronic unease, is one of the best methods to catch and escape the beginnings of an error chain. By continually observing the effects of actions taken and adjusting the plan or mental model as a result, watch standers stay fully engaged in the watch and ahead of the problem.[2]

A certain amount of stress can be helpful to the watch stander. When engaged in a task that does not result in a high-enough level of cognitive burden, people become bored. Bored watch standers become lazy regarding practicing OODA looping, and things get missed. Miss enough things and an accident will happen. The level of cognitive burden needed to prevent boredom is different for each person.

One way to prevent boredom is to take in only the information needed right now. Seeking information that will not be used in a timely fashion can result in a "who cares" attitude. Limiting information allows for only the information needed to be observed, focusing the OODA loop. New information will be just that—new. New information will cause a continuing refinement of the mental model when reinforced by a healthy amount of chronic unease.

ELECTRONIC INFORMATION SYSTEMS

The best way to observe the information needed for right now is to set up the navigation information screens so that only what is needed is actually displayed. Offsetting the position of the ship on the screen opposite the direction of travel or course shifts the display so that more distance ahead can be seen while staying at a lower range. This makes considerable sense for a high-speed vessel, since increased speed reduces the likelihood of being overtaken. Integrated bridge system screen presentations should always be set to "North UP." For ECDIS, this will keep the chart looking like charts have always looked. Having a north-up radar better matches the ECDIS screen, which is critical when quickly sharing information concerning on-screen vectors. Each individual system will have its own procedure for offsetting.

Since ECDIS screens should always be set to north up, they are essentially true presentations. Setting the vectors to "True" will keep the reference for the information presented on the screen the same. The vector time/length should be set to six minutes to allow watch standers to use the six-minute rule.

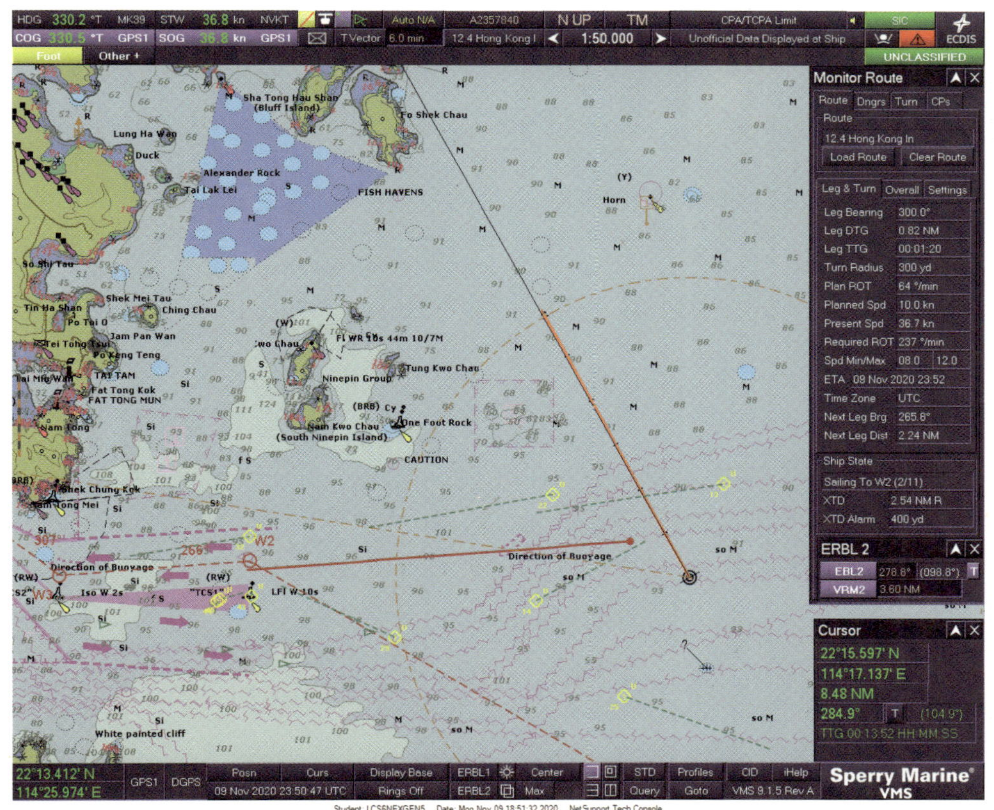

Correctly offset ECDIS screen anticipating course change to the west. *Courtesy of US Navy*

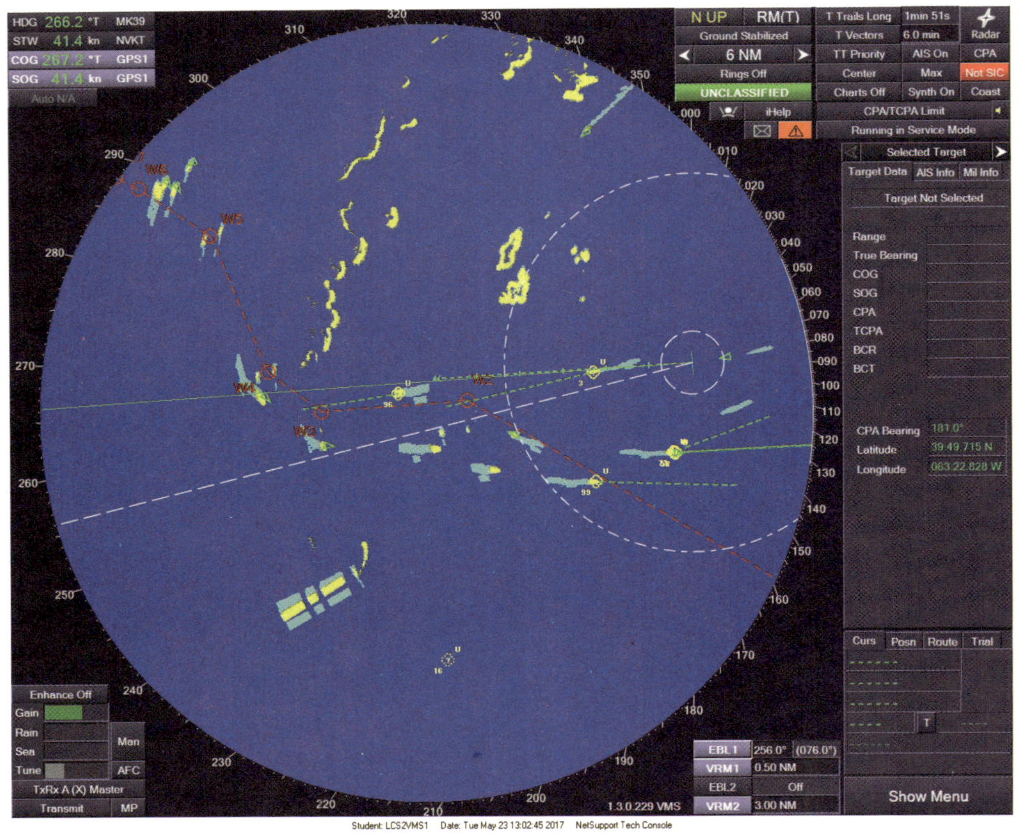

Correctly ranged ARPA screen offset to east, maximizing look ahead while sailing west. *Courtesy of US Navy*

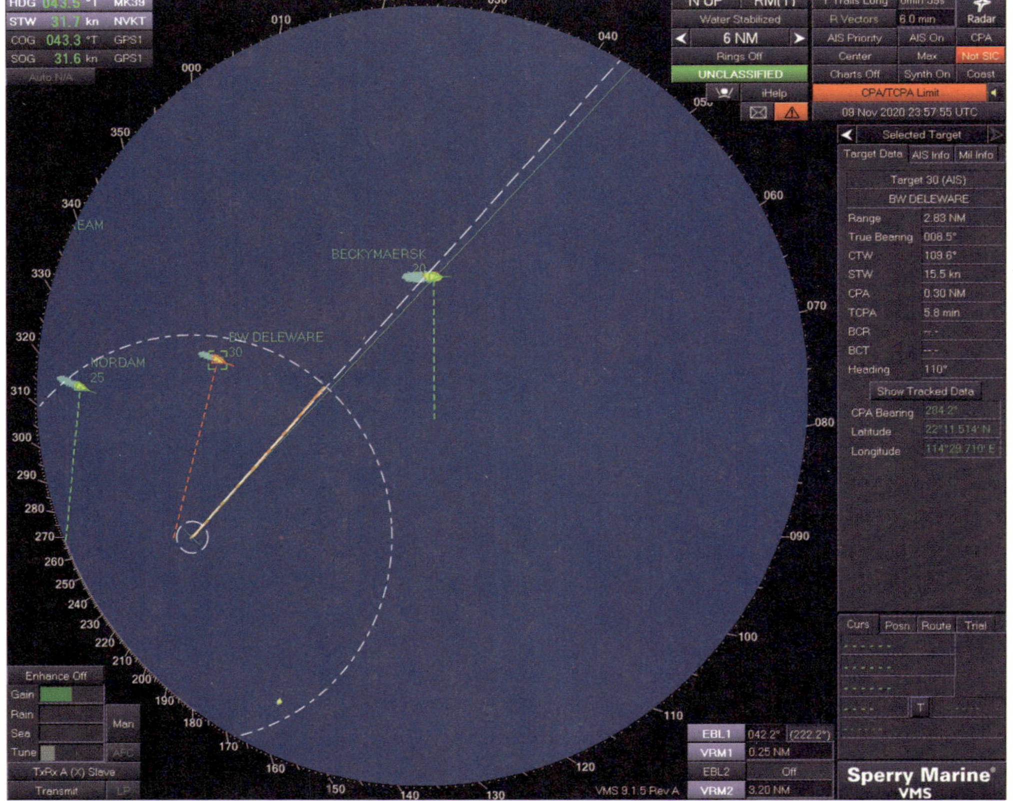

All relative vectors are outside of the safety bubble. While neither vessel will collide with us CPA distance is too close for safety. CPA distance can be further opened by increasing speed.

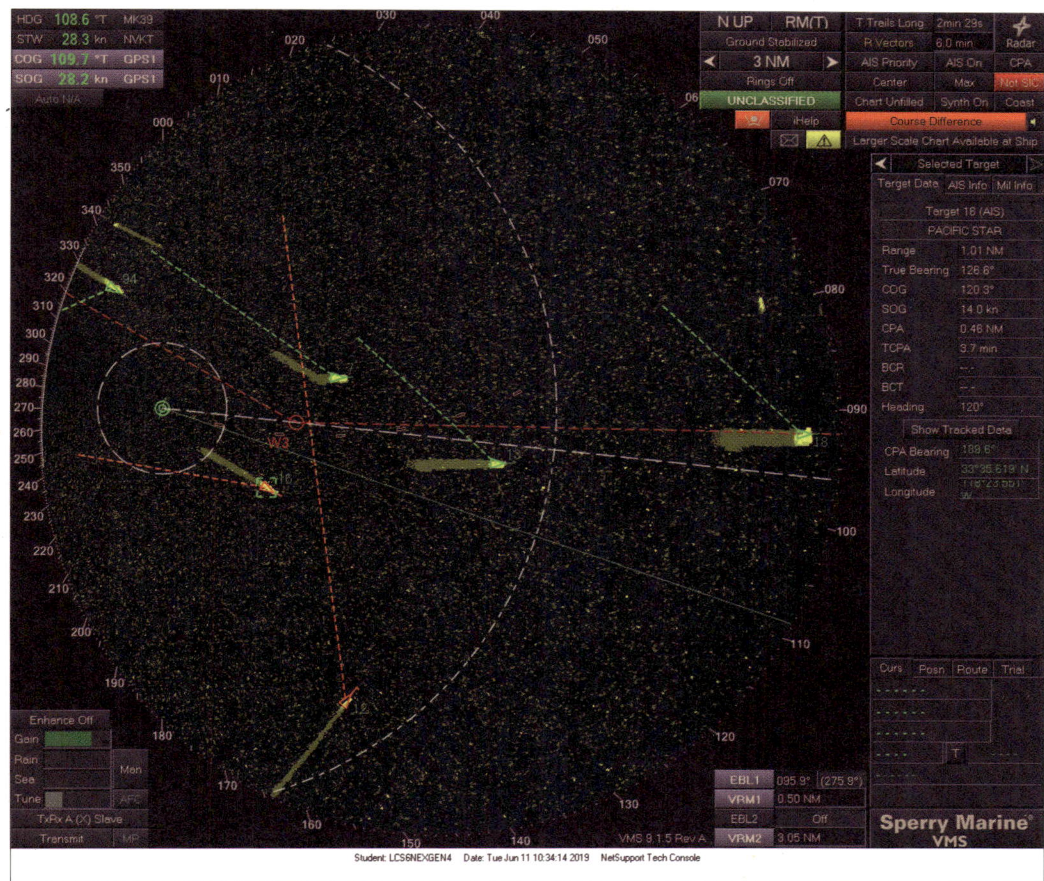

Unsafe relative vector in safety bubble. Come left toward track, but be mindful of vessel broad to port and vessel well off the starboard bow. *Courtesy of US Navy*

The ownship true vector can also be used to help the watch stander determine the maximum safe speed. The screen primarily being used by the captain should ideally present one to two vector lengths of information. Assuming a proper offset, the display scale will be correct when an entire six-minute vector can be seen, plus part or all of another vector. The navigator's screen should display two to three vector lengths. If at any time the proper number of vectors cannot be seen and the offset is correct, vessel speed does not match the look ahead provided by the screen. In this situation, either speed needs to be reduced or the range scale increased to match the speed. Not being able to see the full vector presentation is very similar to a car driving at high speeds on a dark night, with only the fog or parking lights on. The same setup and thought process can be applied to radar offset and vector use, except for using only true vectors. In short, keep appropriate range up before you speed up.

Proper radar use requires switching between true and relative vectors. Both vector presentations have their advantages. A preference for one or the other largely depends on the background and comfort of each individual watch stander. The important thing to remember is that minimum CPA situations require switching to the other vector mode to gain full situational awareness to build a complete mental model.

There seems to be a generational difference in preference for true versus relative vectors. Older sailors tend to prefer relative vectors, while younger sailors prefer true vectors. Both groups will say that "they think better" in their preferred vector format. One theory to explain this is video games and technology. Younger sailors have been exposed to true presentations and vectors from very young ages and may well have an easier time thinking with true vectors. Older sailors may have been taught to use relative vectors in professional training. For this reason it is best to let each watch stander use their equipment as works best for them, provided screens are optimized for easy and efficient information flow.

Both ECDIS and ARPA systems have the ability to display electronic bearing lines (EBLs) and variable range markers (VRMs). These can be used in both a passive and active way to help the watch stander solve the navigation problem.

For radar, a VRM should be centered on the ownship and set to a minimum distance that any other vessel shall not pass inside of, to create an on-screen safety bubble. The recommendation for this minimum distance is 0.25 NM, but the distance can be larger to suit individual preferences. The largest ships approach 1,500 ft. in length, which is just under 0.25 NM. When ARPA is tracking a target there is no way to know the exact spot being tracked. Is the ARPA following the stem, forward wind break, tallest container stack, the deckhouse, or the engine stack? Given this reality setting, the VRM under 0.25 NM would not allow for even the narrowest gap in a worst-case scenario. Displaying an ownship-centered VRM set to 0.25 NM allows for a very quick check of "are we actually going to hit each other." If any part of a contact's relative vector passes through this circle, then both ships will be well inside extremis distances in less than six minutes. This VRM safety bubble works only with relative vectors; true vectors require a different set of logic.

PREDICTED TRUE VECTOR

True vectors show not only where our ship will be in six minutes, but also where any other ship, with a vector, will be in six minutes. Comparing the relationships of the various true vectors can help build a mental model not only of just how will two or more vessels pass each other, but also where they are all trying to go. This second piece of information can be very important when navigating near a coast or in bays. In these areas, vessels tend to be less able to maneuver freely due to draft, channel, lane, and weather conditions. Understanding where another ship is trying to get to can have a significant impact on how the navigation problem is solved.

To compare vectors to solve the puzzle, watch standers should use the predicted true vector, or PTV. First an ownship-centered VRM is displayed, set to the current six-minute distance at one-tenth the current speed. This circle represents where the ship could be in six minutes after any course alteration. Next, an ownship-centered EBL is set to the current course being steered. When the six-minute vector becomes involved (near to) with another ship's true vector, risk of collision most probably exists. Often these vectors will cross. If the length of the vectors could be quickly and accurately measured, the difference in length of the vectors after the crossing would equal CPA distance. This is not really a practical way of using the presentation. However, most people are capable of seeing if the vectors are of equal length after the crossing point. If this is the case, action needs to be taken. Manipulating the EBL will show how changing course either right or left will affect the crossing or CPA distance. If a course change is not sufficient on its own, or not possible, then adjusting the VRM closer shows the effects of reducing speed. If an increase in speed is possible and will happen quick enough to be effective, the VRM can be extended to show how increasing speed will change the situation. In a sense, a properly used PTV allows the bridge crew to graphically trial-maneuver in real time. Using the PTV in lieu of trial maneuvers is considerably faster and does not carry the potential of accidentally leaving the ARPA in trial mode, which has been the cause of more than a few ARPA-induced collisions.

The PTV has another use beyond traffic management. When displayed on the ECDIS, it can help the navigator pick their way not only through traffic, but other navigation features. Aligning the EBL with the intended course prior to a waypoint or course change will show if an early turn is available, or how long a turn can be held off. In this way it is possible to "think" with the PTV, and as such it should be actively used on the ECDIS during the course of a watch. The truly interesting thing about the PTV is that it can be used on any ARPA or ECDIS at any speed over ground. It is not a solely high-speed concept.

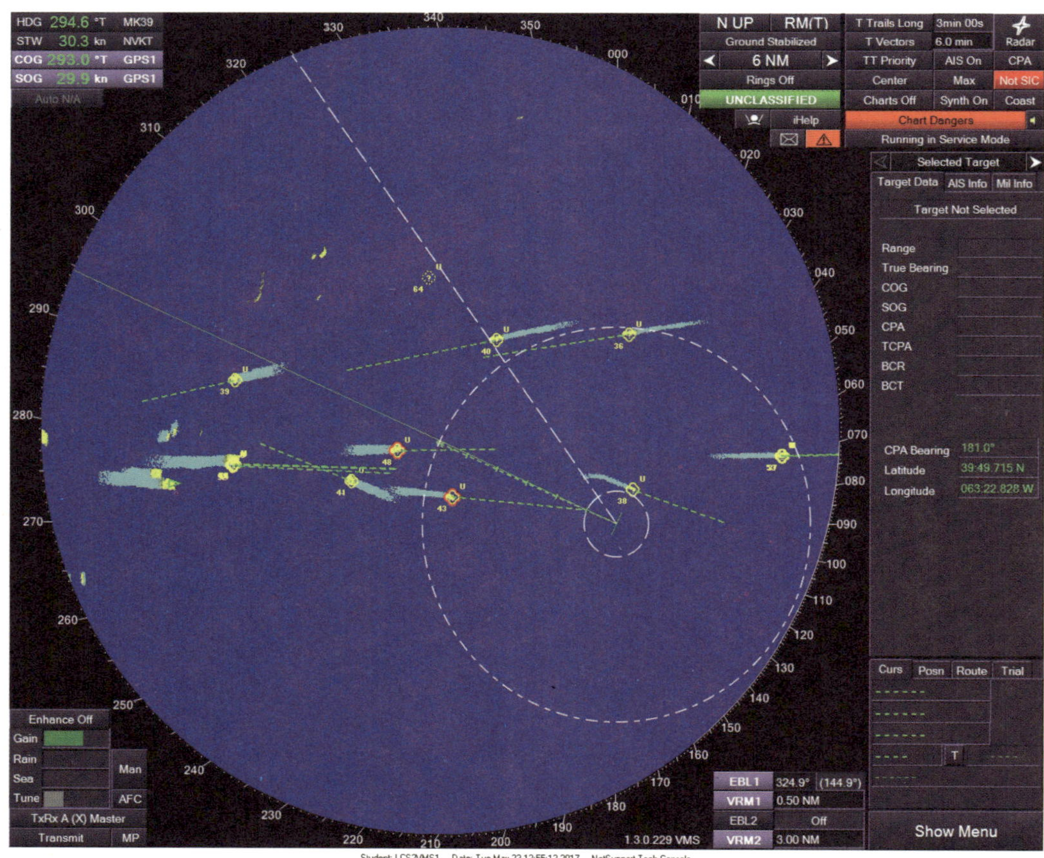

PTV showing potentially safe course assuming we turn to the west before reaching the tip of the PTV. *Courtesy of US Navy*

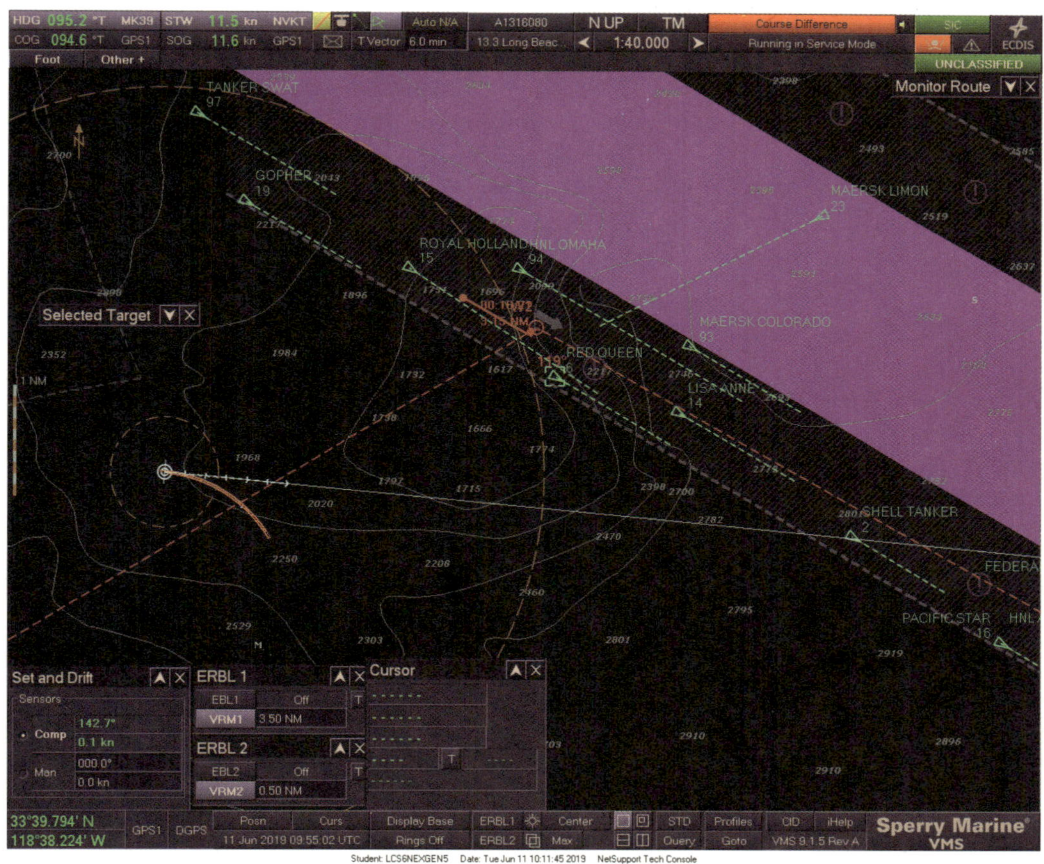

PTV showing that it is safe to increase speed from 11.5 to 35 knots. There are no vector tips near the tip of our proposed vector. *Courtesy of US Navy*

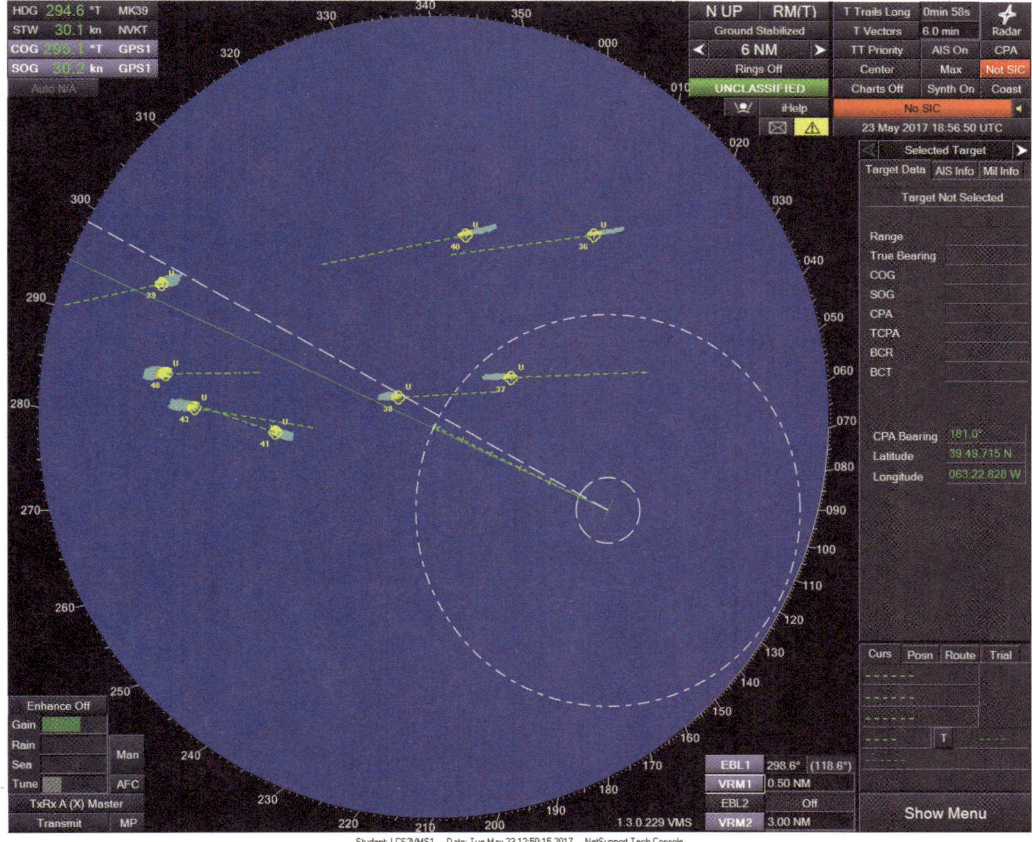

Current course and speed are safe. No true vector tips near ours. However, this may need to be reevaluated in the future due to targets off the port bow. *Courtesy of US Navy*

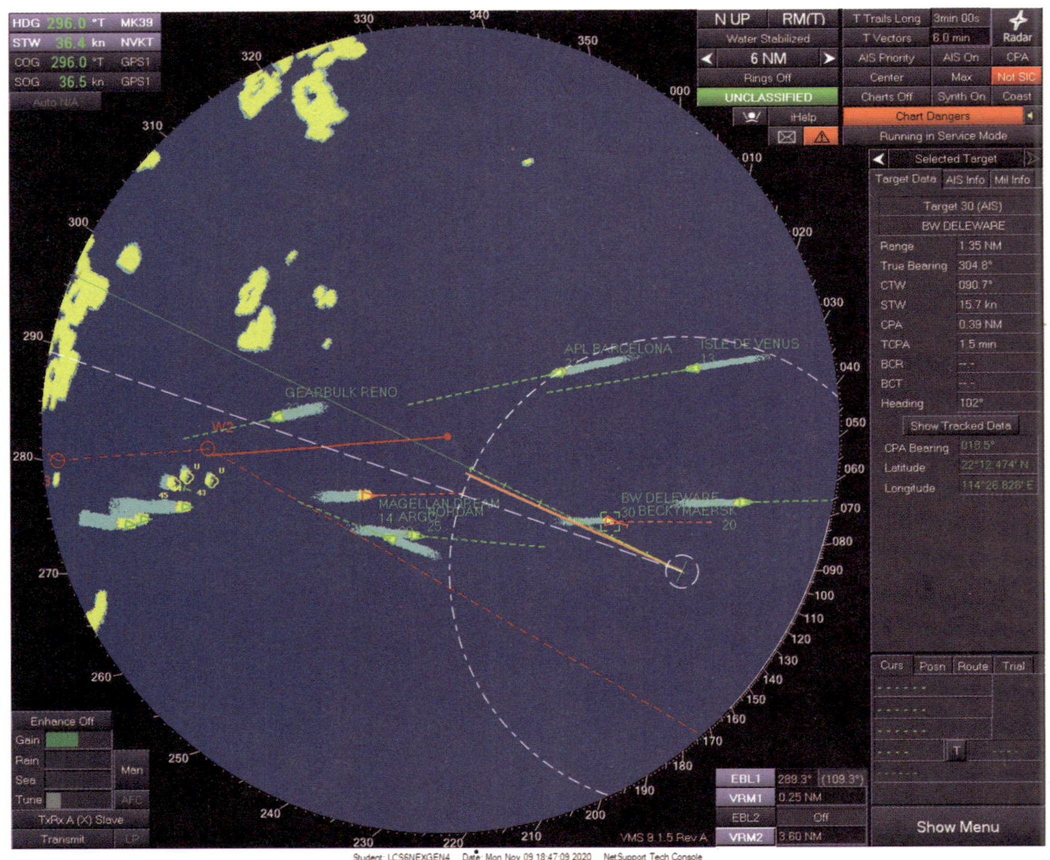

A few minutes later, there is now a need to consider altering course. Once the contact broad to starboard passes, a hole opens to the north. *Courtesy of US Navy*

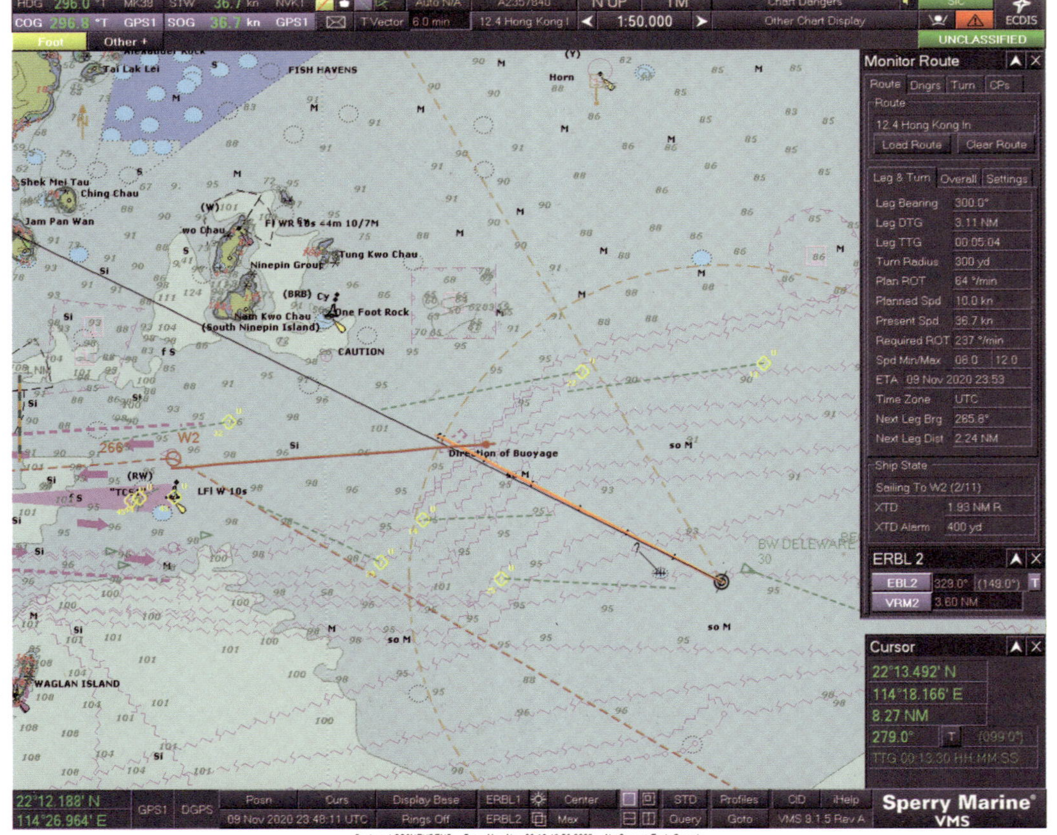

ECDIS presentation showing PTV to NNW. No vector tip are near the PTV tip. Our ship will pass ahead of the ship to the right of the PTV by about one-and-a-half minutes, but we will turn west well before that. *Courtesy of US Navy*

ECDIS AND ARPA SCREEN SETUP

Another tool that allows for quick, efficient positioning is the radar overlay. Many ECDIS are capable of displaying the radar image directly over the chart. This can be very useful when entering a new port where the accuracy of the chart image has not been verified. A variance between the radar image and the chart that cannot be explained by features drying at low tide, which should be so marked, immediately requires additional scrutiny of the vessel's position.

The voyage route should be entered and displayed on the ECDIS. Doing this makes it very easy to quickly determine if the current cross-track error is significant and potentially indicates a growing chain of errors. Along with the route displayed graphically on the chart and ARPA, a window showing the actual route-monitoring information should also be displayed where it is visible to the members of the bridge team while in their chairs. This can be on a centrally positioned screen or displayed on each personal screen.

Finally, a properly setup display will show true trails on the ARPA screen. Properly tuning a radar in declining weather conditions is a balancing act and will not always produce the clearest picture. Displaying true trails will help alleviate this problem. When trails are set to long and are not frequently cleared away by changing scale or offset, they will show the history of every target the radar is trying to display through the clutter. In this way, all targets can be observed. Targets with longer trails are going faster and need more attention. Buoys and anchored ships can have trails, but these will be obviously shorter when compared to even the slowest underway contact.

In short, a properly setup display will have both the radar and ECDIS in north up with true trails. The offset will be set to maximize the look ahead while keeping the range low to increase the clarity of the chart or radar picture. The route will be displayed both on the chart and in the route monitor table. The ECDIS will have the radar overlay displayed. The vector time/length will be six minutes, with true vectors selected for ECDIS. ARPA vectors will also be six minutes but be set to either relative or true depending on user preferences. The ARPA will have a safety bubble centered on ownship. The PTV will be set up on both the ECDIS and on the ARPA, since it is useful on both screens. A bridge that is set up in this way will lend itself to being useful by one's merely glancing down.

This glance-down-capable setup will require the operator only to adjust offset, range, and the PTV, any of which can be done within the seconds that the captain is allowed to be looking down without using the eyes-up/eyes-down procedure. This serves to reduce cognitive burden, increase the building and sharing of the mental model, and keep stress levels appropriate.

All these concepts will aid the bridge team in making decisions that allow them to continually and consistently put the ship in the middle of the good water—which does not mean being in the middle of a slip, channel, or lane. It means being in the best spot between all the currently pressing dangers. This balances the needs of safety and prudence with the requirement of getting to the other side of whichever body of water is being crossed.

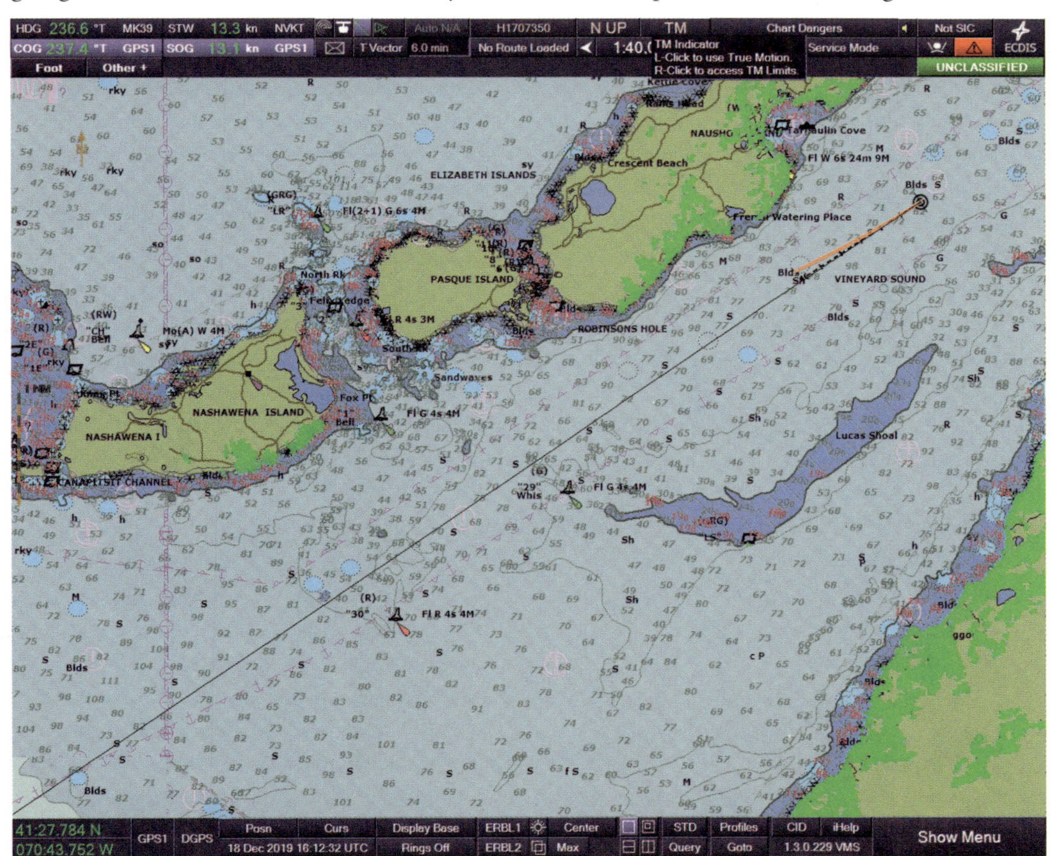

Radar return overlay on the ECDIS screen. *Courtesy of US Navy*

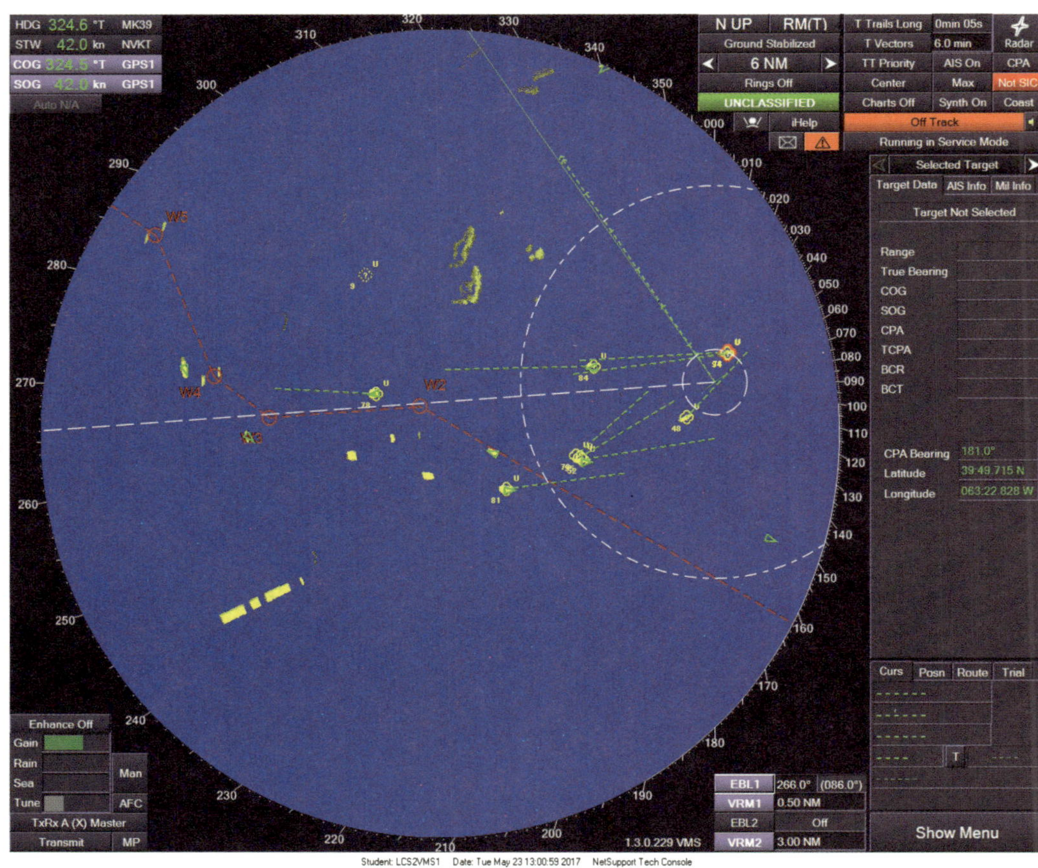

Voyage route displayed on the ARPA screen. *Courtesy of US Navy*

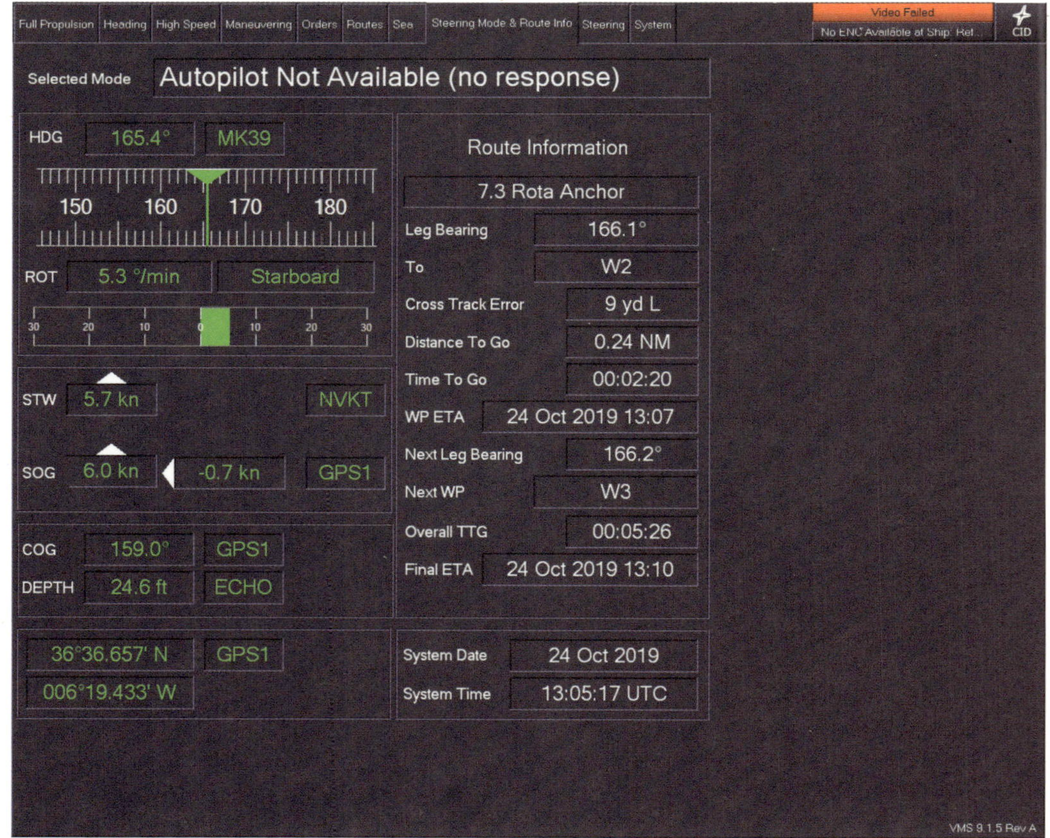

Route information displayed on a conning information screen. *Courtesy of US Navy*

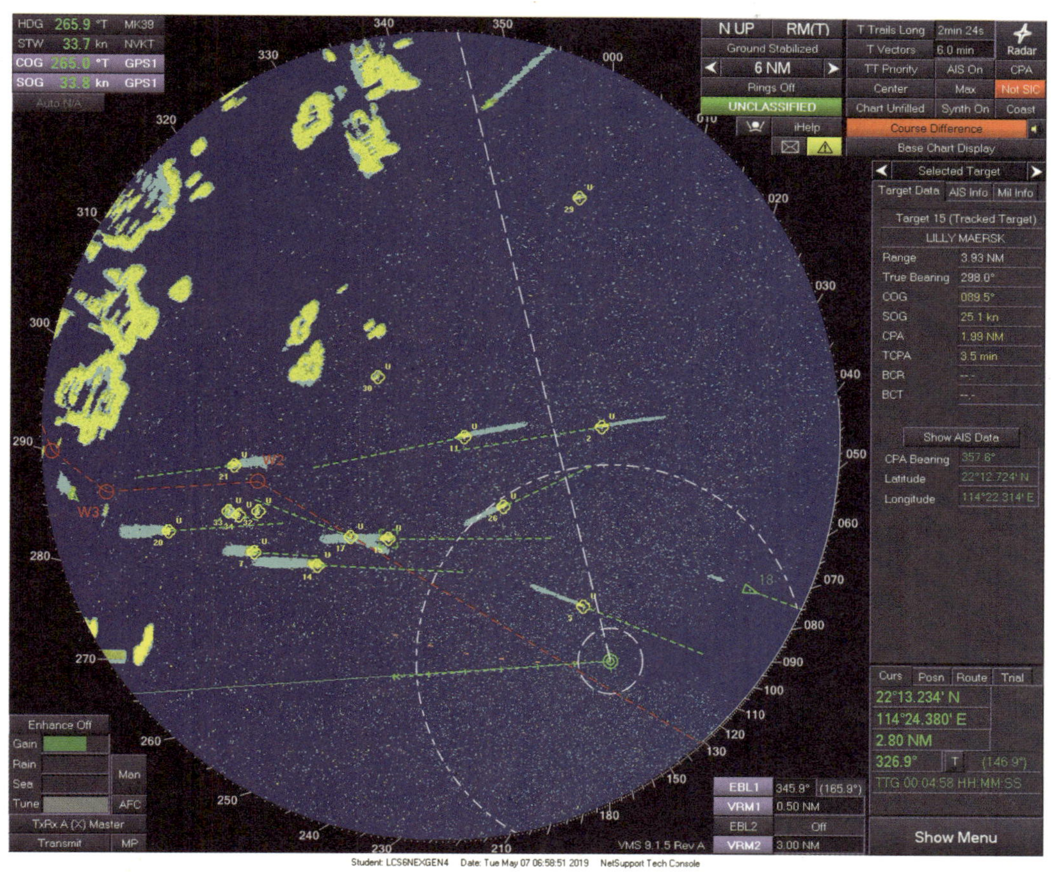

Even though it's heavily speckled, this is still a good picture. The radar is cutting sharp edges on the land and surrounding ships. Avoid overtuning the screen or targets may be missed. *Courtesy of US Navy*

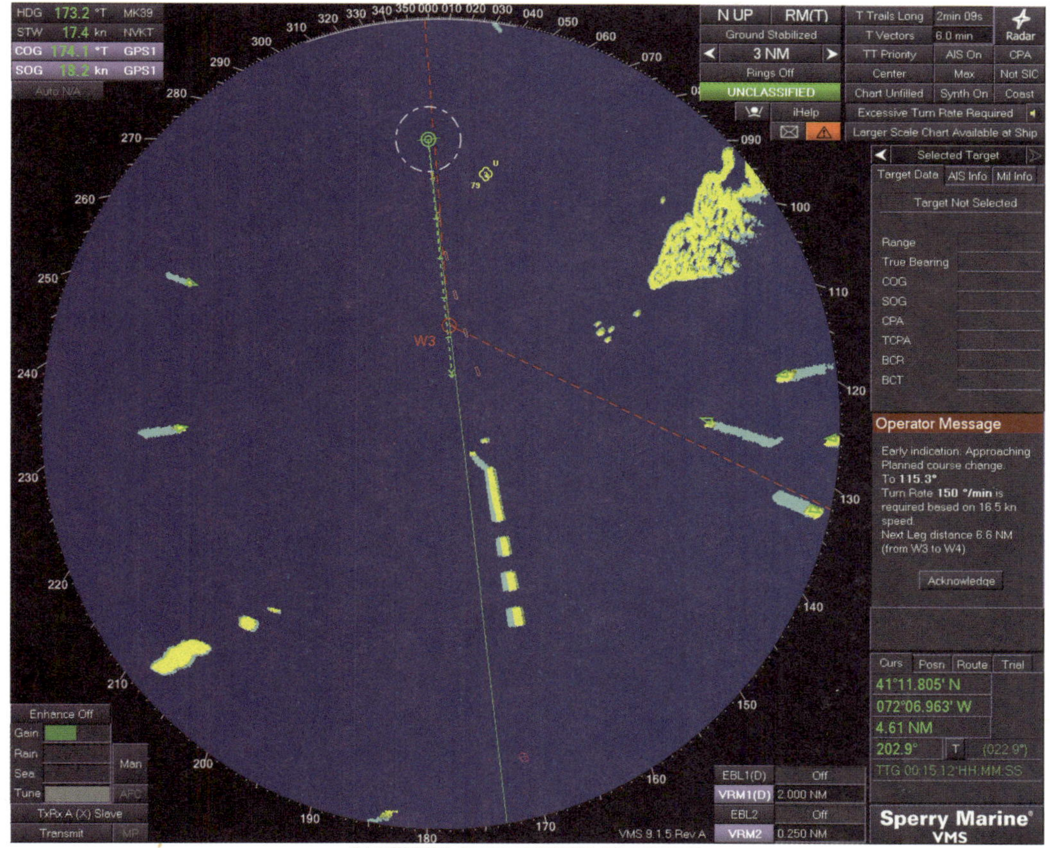

Clear true trails in good weather. *Courtesy of US Navy*

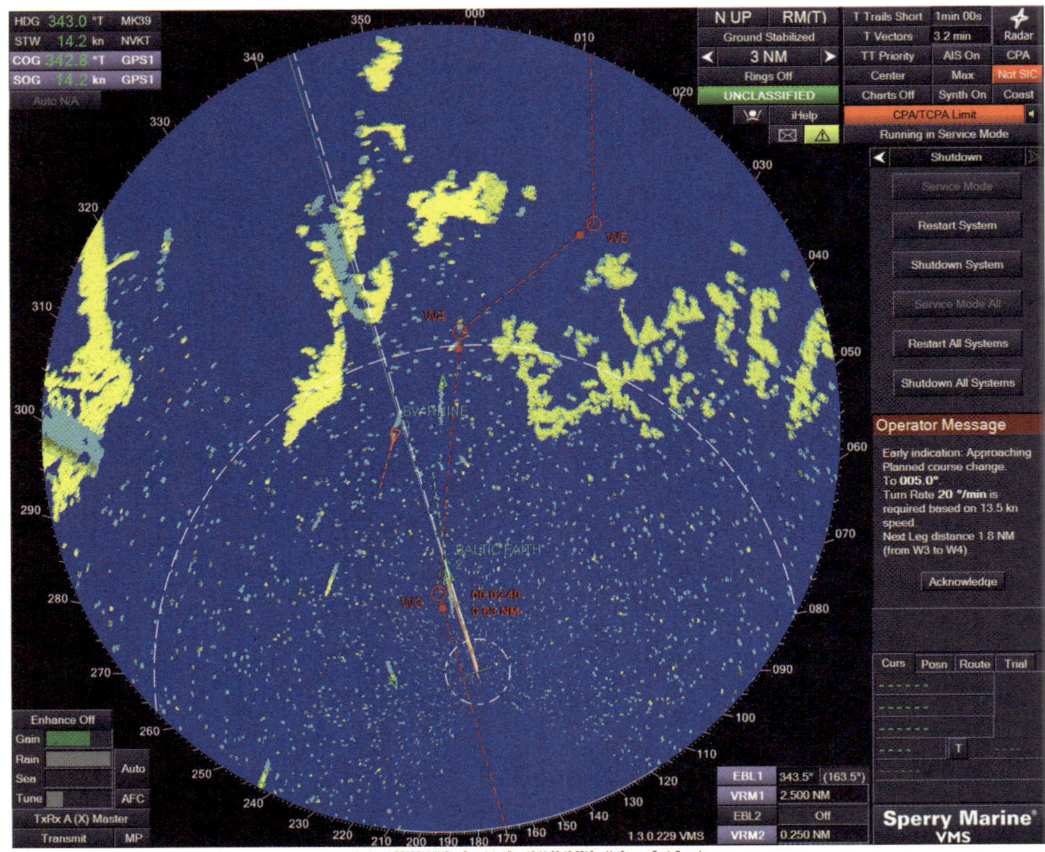

True trails are still clear and distinct even though the picture is degraded by sea clutter. *Courtesy of US Navy*

CHAPTER FIVE

OTHER CONSIDERATIONS

AUTOPILOT

Autopilot is a critical tool for the watch stander. Proper use will allow a higher degree of situational awareness and a lower level of overall mental stress. It can be hard to keep a water jet vessel on track while trying to accomplish other parts of a navigational watch, especially at higher speeds.

The various settings and how they need to be managed under differing speeds and weather conditions need to be fully and properly understood for the safe navigation of the ship, making it critical that all watch standers be familiar with the operating manual or be instructed not to change settings without approval.

Use of the autopilot outside the recommended safe distance from other navigational dangers takes an additional task off the watch stander, allowing more on the navigation and traffic picture. This will tend to reduce the mental stress on the bridge team, providing for better BRM and situational awareness.

When the ship is going fast, autopilot is the preferred method of steering, with the added caveat that CPAs must be adjusted accordingly. If the CPAs cannot be adjusted and continuing at high speed is required, then it is preferable to switch back to hand steering.

Making turns in autopilot rather than by hand steering must take into consideration the off-course alarm and potential for the autopilot to drop out. Water jet vessels are able to turn very quickly and may outpace the off-course alarm or the maximum rate of turn setting. Either of these conditions could cause the autopilot to alarm and possibly drop out, setting the vessel into danger.

Track mode is another type of autopilot that does more than hold heading. Track mode is capable of following a voyage plan, including speed changes. To do this, it takes in more points of information than autopilot and relies on a correctly entered voyage plan with proper settings and limits. Use of track mode should be kept to open-ocean, sea buoy–to–sea buoy transits. While the technology exists to allow for dock-to-dock passages, there are too many settings both in the autopilot itself and in the ECDIS that if incorrect could cause an autopilot dropout or ECDIS-induced grounding.

Use of autopilot in pilotage waters will lead to the bridge team spending as much time paying attention to the autopilot as they would to actually driving the ship instead. Except when the autopilot is steering, there will be an increase in the time needed to figure out why the ship is standing into danger, compared to having the helm in hand and knowing how all the forces have been acting on the ship up to that point. This represents a loss rather than improvement in situational awareness.

When operating well away from navigational dangers, such as shoal water, land, or other vessels, it is in the best interests of the bridge team to make use of autopilot. A properly configured autopilot will keep the ship on course, although distance to danger and the reaction time needed to switch into manual steering need to be kept in mind when deciding to make use of autopilot. Proper use allows the watch to focus more attention on the navigation picture, working toward getting ahead of the problem. The very fact that the vessel is operating at higher speeds implies that there is sufficient distance from shore and other dangers; otherwise the damage from the ship's wake could be enormous.

Restricted maneuvering situations, such as channel transits or proceeding to an anchorage, require a high degree of concentration on the part of the watch stander. These situations also reduce the available reaction time needed to recognize and switch off the autopilot in the case of an emergency. The issue with using autopilot close to shore has little to do with the ability of the autopilot to hold the desired heading at slow speeds. It has more to do with the inability of the watch team to recognize that autopilot is unknowingly standing the ship into danger. Reluctance to turn the autopilot off can prevent the error chain from being broken.

To many, it may seem that with the improved handling inherent to water jet ships that autopilot can and should be used as a dedicated helmsman. This is poor Bridge Resource Management. Putting a computer in charge of this critical duty during a transit will not reduce the required number of bodies on the bridge; someone will still need to watch the autopilot. As the distance to danger is reduced, the need to watch the autopilot will take all of that person's focus, taking their eyes out of the window and forcing them to stare at the autopilot screen. No reasonable mariner will agree that this method of vessel control is better than steering the vessel manually while looking out the window.

Any comparisons to drones or self-driving cars are false ones. Those systems take in much-larger swaths of information and make decisions. Autopilot is capable only of determining if the vessel is still on course or track, and making adjustments to correct. There is no additional information entering the process or higher-level considerations.

An additional impact of using the autopilot in these situations is skills atrophy: when a skill is not regularly practiced, an individual's ability in that area reduces. Not taking the time to practice hand steering the ship when it is not required will add to the difficulty of dealing with the situation when autopilot fails. Knowing that skills have been weakened over time may give a watch stander pause when the need arises to hand steer. This lack of confidence will also add to the complexity of the situation. Skills atrophy and loss of confidence are both easily prevented by simply steering the ship by hand when required instead of shying away and using autopilot as a crutch.

TUGS AND THEIR EFFECTIVE USE

Generally, tugs and water jet ships do not mix well. With the need of water jet ships to be efficient, they must be built light. In an effort to reduce the weight of the hull, materials such as aluminum and fiberglass are typically used. It is possible that in the future, carbon fiber could be used, but this will require a significant reduction in manufacturing costs.

Since the hulls of these ships are built with lightweight materials and framing, there is the potential to cause damage to the hull when bringing a tug in contact with the skin of the ship. While the global structure of a water jet ship may be very strong, the skin of the ship is often much thinner than found on steel ships. This requires special attention to avoiding point-loading any one section of the hull. Many tugs use chains to attach their fenders. If these chains are caught between the fender and the hull of the ship, a puncture or gouge is quite possible.

Another difficulty in making effective use of tug boats is the power available to the tug operator. The power of modern tugs is matched to vessels approaching and even exceeding 100,000 deadweight tonnage (dwt). Tugs with this kind of horsepower can easily overcome the various environmental forces that act on the ship, and even the controllable forces the ship can develop on its own. This is potentially a dangerous situation for the operator of a water jet ship. If a tug operator is not completely familiar with water jet ships, they can very easily turn a manageable docking evolution into a disaster.

Complicating this issue are the safe working loads of the mooring gear on the deck of the water jet ship. The capstans and bitts are matched to the strength of the lines needed to moor the ship. Given the lightness of these ships, safe working loads around 20 tons are quite common. Using this gear with tugs capable of bollard pull exceeding 60 tons invites problems.

One possible solution is to have a small boat serve in place of a tug. Water jet ships are capable of developing significant controllable forces. When the environmental forces exceed the ability of the ship, often only a slight push is needed. Having a moderately powered rigid-hulled inflatable boat (RHIB) or other well-fendered small boat nose into the bow and push on the lee side will be enough to set things right. In a sense the small boat will act like a bow thruster to help hold the bow up into the wind. If the ship is being set down on the pier, the same small boat can pull on a ship's line to keep the lateral speed controlled.

When using a small boat or tug in this way, the maneuvering setup should be adjusted to match the description of working with a thruster. Otherwise the ship will not be maneuvered efficiently, increasing the potential for an accident.

When there is no way around using tugs, try to avoid taking their line, since it is matched to the tug's bollard pull. Instead, offer a ship's line by sending down an eye and then making the line fast to the strongest mooring feature nearest to the chock. If that is not acceptable, another solution is to have on hand a short length of mooring line that can be passed through the eye of the tug line and then back through the chock and laid down so both eyes are on the bit. This short length of mooring line works as a weak link parting before the safe working load of the mooring equipment is exceeded. When wires are used to work the ship, these short pennants have the added benefit of keeping the wire from coming in contact with any part of the ship. If the vessel's mooring tackle is constructed from aluminum, significant damage may occur from contact with these wires.

EMERGENCY MANEUVERS

MAN-OVERBOARD MANEUVER

Every ship has a preplanned response for dealing with a man-overboard emergency. Most ships make use of the Williamson, Anderson, or Round turn. All these turns allow for the ship to return to a specific point along her track without the need for twisting to come about to the reciprocal heading. The trade-off is distance away from the victim. The thrust vectoring available to water jet ships eliminates the need for this trade-off. When operating at speed, it is entirely possible that a Williamson turn will take the ship three times the distance from the victim or more when compared to stopping the vessel and twisting in place.

The stop-and-twist maneuver, similar to a "J turn," reduces the distance from the victim as well as the potential of cavitation, which is present during all the previously mentioned turns due to the length of time the buckets must be held hard over.

The stop-and-twist maneuver for a vessel with multiple jets is conducted as follows:

1: Check behind the vessel to ensure there is no one close astern, and mark the vessel's heading.

2: All buckets should be full astern (this will cause the backing plates to close, generating reverse thrust).

3: When the backing plates begin to close, set all buckets hard over to start the twist. Going hard over at this point will not reduce the required stopping distance. It will start rotating the bow around as the ship comes to a stop. This will reduce the time of the maneuver when compared to starting the twist once fully stopped.

4: While backing with all jets astern, select the steering mode that allows for split-plant maneuvering and continue backing until going approximately 4 knots astern. While this much sternway may seem excessive, it will allow for higher initial ahead rpm, which serves to dramatically increase the twist's rate of turn by working against the vessel's inertia.

5: When sufficient sternway has been achieved, come ahead to maximum cavitation rpm (ZSTW) on one side of the ship and shift the jet toe angle for those jets. This will put the ship into a high-power toe-in or toe-out spin. Adjust ahead rpm so that the vessel does not cross her own wake. Keeping speed regulated this way will ensure that when the head has come around 180 degrees, the victim will be either dead ahead or fine on the bow.

6: If it is desired to come ahead before coming to the reciprocal heading, keep twisting until the victim / reciprocal heading is less than broad off the bow before coming ahead to maximum cavitation rpm (ZSTW) on all jets.

7: When the decision to come ahead is made, bring all jets to the ahead maximum cavitation rpm (ZSTW) and proceed back to the victim, taking care not to steer directly at them.

8: As the ship closes the victim, reduce speed by splitting the plant with the backing jets on the side of the rescue boat. Use the ahead rpm to control speed and heading to create safe launching conditions for the rescue boat. Take care to stay a minimum safe distance from the person so that they are not placed at risk of being sucked underwater and into the jet inlets while the boat is being launched.

9: Maintain a safe distance and continue to safely navigate the ship while waiting for the rescue boat to return. Recovery of the rescue boat should be a process very similar to launching.

This maneuver can also be accomplished by a boat with a single jet. All steps are the same except for splitting the plant and twisting. Here the single-jet drive would just back until 4 knots astern, then shift the bucket ahead and from one side of the keel to the other while keeping from running over her own wake.

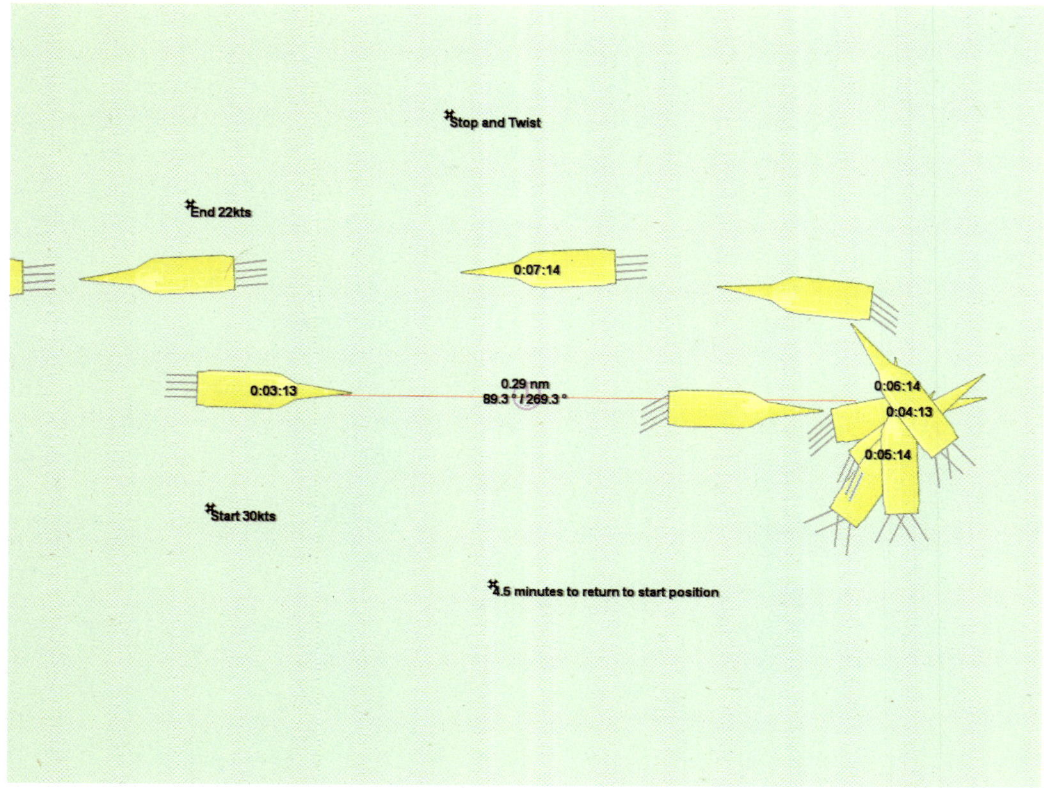

Back-and-twist recovery maneuver. *Courtesy of US Navy*

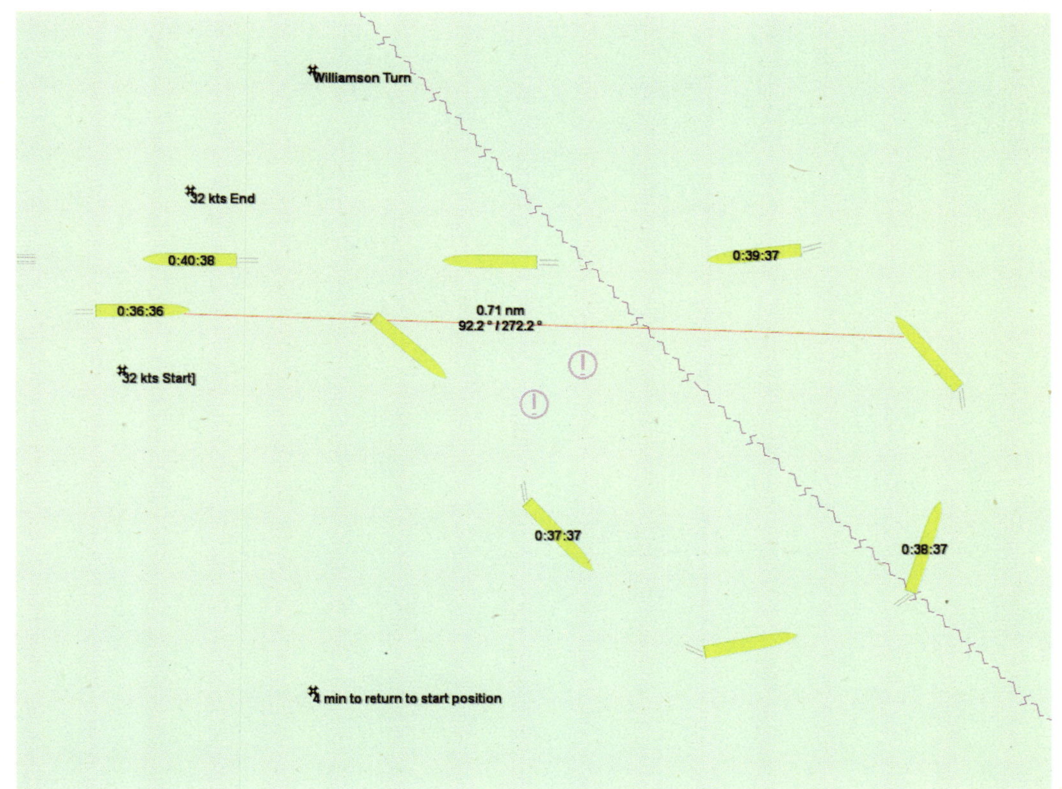

Williamson turn recovery maneuver. *Courtesy of US Navy*

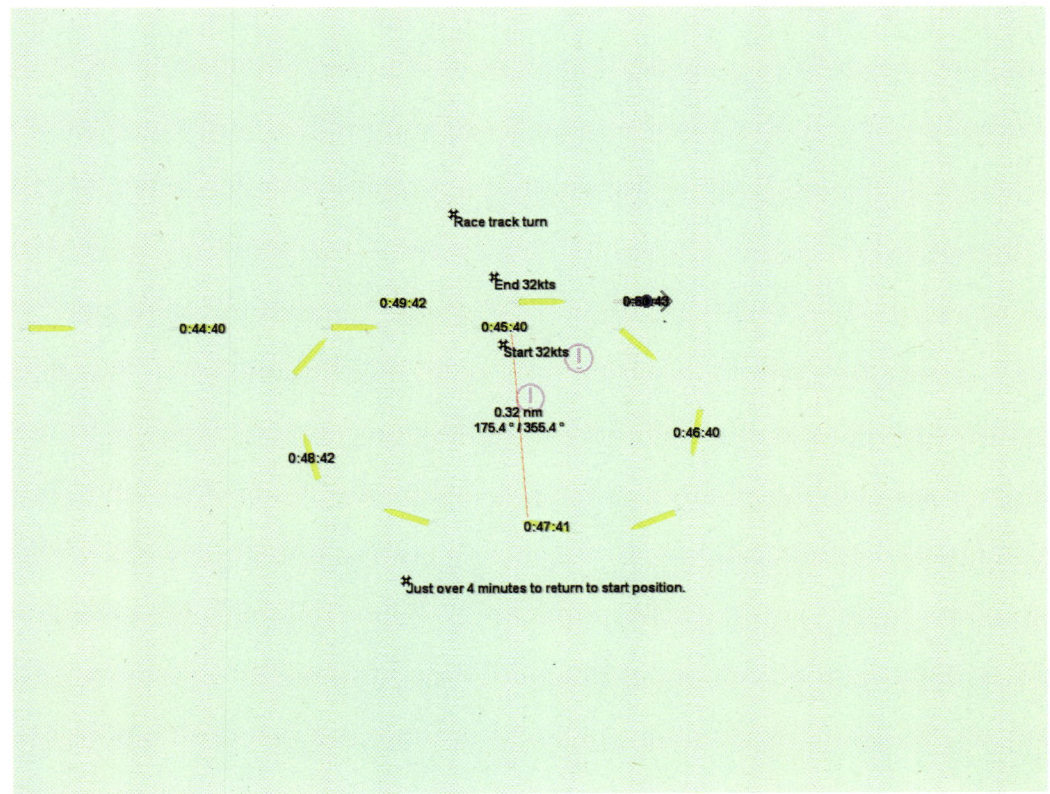

Anderson turn recovery maneuver. *Courtesy of US Navy*

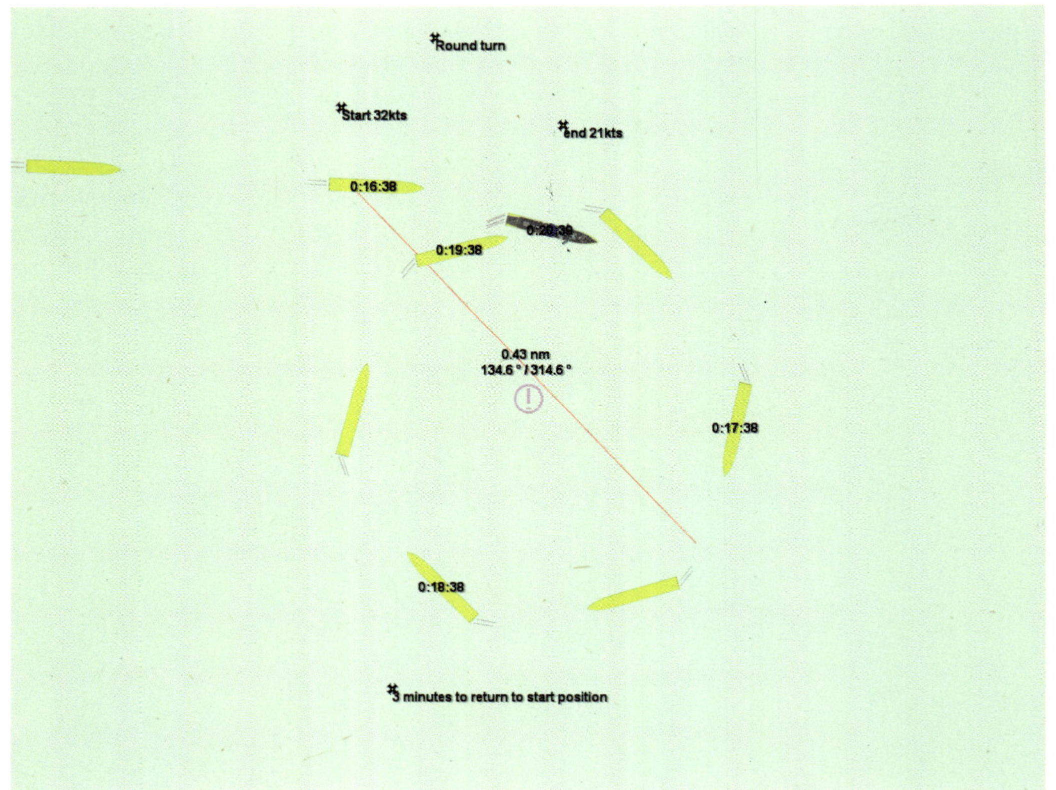

Round Turn recovery maneuver. *Courtesy of US Navy*

APPENDIX

NOTES

1. Bill Lyons can often be heard making these points to students as he teaches.
2. The OODA loop was developed by Col. John Boyd to help fighter pilots obtain higher levels of situational awareness while in a dogfight.

GLOSSARY

active jet: The jet used for steering control while in split plant. The ahead jet will be active when going ahead or walking. The astern jet will be active when the ship is backing.

actual speed: The current speed of the vessel as measured either over the ground or through the water

Anderson turn: A turn that uses specific rudder angles and wheel-over points to bring a ship back on her reciprocal course for recovery of a man overboard; it does not take into account cavitation management.

angle of intersection: The angle that the divergent wake and transverse wake cross. An angle of 90 degrees causes the greatest increase in the power of the resulting wake.

ARPA (Automatic Radar Plotting Aid): A computer system paired with radar that solves a target's relative and true motion for the watch stander. Used to predict time to closest point of approach (TCPA) and closest point of approach (CPA).

ARPA-induced collision: A collision where a major factor was the improper use or interpretation of ARPA data by the watch stander

autopilot: A computer system that measures the difference between the ordered course and the actual course. Then, operating within the constraints of the various control settings, it uses the vessel's steering system to return to the ordered course.

axial-flow impeller: An impeller design that produces thrust by pushing water along the axis of flow into the thrust nozzle

backflushing: Process where water is forced through the impeller and tunnel opposite of normal flow to push an obstruction out the inlet of the tunnel

backflushing procedure: The specific steps required to allow for backflushing to work. This is specific to each ship and is good to have on hand in checklist form at the conning station.

backing plate: Part of the bucket that closes to direct the jet wash so that the ship is thrust astern. Can be combined with the reverse deflector.

backup control system: An redundant system that allows the operator to control the buckets and engines without making use of either the steering or engine control systems. Similar in function to a non-follow-up system.

blade face: The area of the impeller blade

blade height: The height of the impeller blade as measured from the central hub to the blade tip

blade leading edge: The front edge of the impeller blade that separates the flow to the front versus the backside of the blade face

blade length: The length of the impeller blade as measured from the leading edge to the trailing edge

blade tip: The part of the impeller blade that is farthest from and parallel to the hub

blade trailing edge: The back edge of the blade where the flows from the front and back of the blade face come together

bollard pull: The amount of pulling force a tugboat can produce, typically measured in tons

bottom effects: The different ways that pressure waves formed from a vessel traveling through the water impact shore or bottom features and then reflect and act on the same vessel.

bow thruster: A propulsion device mounted forward that is typically used to assert lateral control on the bow. Most often used in slow-speed-maneuvering situations, such as docking or station keeping.

bow wave: The wave created as the bow of a vessel passes through the water. This wave is easily visible on the surface. This wave can also be referred to as the divergent wave, which, crossing the stern wave at nearly right angles, has the potential to produce damaging waves.

breaking strain: The amount of strain that can be exerted before an object breaks. In this case it refers to the point at which mooring tackle will fail and pull out of the ship's decking.

BRM: Bridge resource management is a process where positive communication tactics and information are shared among all members of the bridge team to effectively run a navigation watch.

bucket: The structure that is mounted to the nozzle and surrounds the exiting jet wash. Used to control steering and backing forces.

catamaran: A multihulled vessel consisting of two full-length hulls in parallel, connected by a bridging structure. The area above the bridge is where passengers or cargo is carried.

cavitation: Vapor bubbles created in the water as its pressure changes while passing through the impeller

cavitation damage: Impact corrosion caused by the vapor bubbles collapsing against the impeller blades or water jet tunnel

cavitation rpm limit: The maximum shaft rpm that can be sustained without creating cavitation on the impeller blades. This limit changes with speed through the water.

cavitation table: A mathematically derived chart that compares speed through the water with shaft rpm against a presolved cavitation line

chart overlay: A limited depiction of a chart that is displayed on top of the radar image presented by the ARPA

chronic unease: A feeling—which is always present—that something is missing from a person's understanding of the world around them. This is not a bad thing, nor is it to be avoided.

clutch: A mechanical device that connects parts of a drivetrain. The clutch allows a prime mover, such as a diesel engine or gas turbine, to idle, thereby not producing any impeller rotation.

cognitive burden: The amount of thought processing currently being expected of a person. Equal situations will not result in the same level of cognitive burden between different people; age, training, experience, and ability are reflected in the level of cognitive burden individuals feel.

combinator: A bridge control that allows the watch stander to control two actions at once. For water jets, this would be jet angle and drivetrain power level.

complacency: Being satisfied with current levels of self performance combined with a lack of situational awareness

conning station: The position on the bridge where the navigation/steering watch is kept. Typically an integrated bridge system is installed at this location.

continuous-operation zone: The area on the cavitation table where no cavitation condition exists for a given shaft rpm and speed through the water

controller pitch propeller: A propeller where the pitch of the blades is variable. This arrangement allows for continuous rotation of the shaft with limited to no effect on the propulsive forces being developed.

crabbing: Proceeding ahead or astern so that the leading end of the ship is upwind/current. This allows the course over ground to align with the channel/slip heading, not the heading of the ship. Often referred to as the angle of crab.

deep water: Water depths where the bow wave is able to rotate under the ship without touching the bottom. This depth will be different for each ship, depending largely on draft.

displacement hull: A hull design where the weight of the hull equals the weight of the water contained in the same volume. These hulls are supported by the water column below the keel.

displacement pump: A pump that is capable of pulling the liquid in that it is trying to pump. These pumps are used when low volumes of high-viscosity liquids are being pumped, and additionally are used when the possibility of pumping air along with the liquid exists, such as stripping pumps on tankers.

divergent wake: The horizontal part of the bow wave that extends across the surface of the water

DIW (dead in the water): A vessel not using propulsive force to control speed or heading

docking station: A station similar to the conning station, but optimized for docking evolutions. Typically these stations have more steering modes available but fewer navigation systems. Often located where the best view of docking is expected, such as the bridge wings, or centrally but aft facing.

ECDIS (Electronic Chart Display Information System): Allows for virtual instant position plotting on a chart, made up of various information layers. These are required on larger oceangoing ships and have largely replaced charts on high-speed vessels.

ECDIS-induced grounding: A grounding where the watch stander misused or misunderstood the information presented on the ECDIS screen

effective wash: The amount of jet wash through the bucket required to maintain control of the vessel

electrohydraulic steering-control system: The combined electronic and hydraulic system used to transmit steering commands from the conning station to the hydraulic steering rams

electronic bearing line (EBL): A function of radar that allows the user to lay bearing lines from ownship, or from a user-defined offset origin

engine control room: A space typically near the vessel's engines that contains the equipment and systems needed for the engineering team to operate the engineering plant

engine control system: A computer-based system used to monitor and control the various engineering systems needed to operate the vessel

engine room management: A process similar to BRM that allows the engineering team to share information among all members to effectively run the engineering plant

error chain: A series of events that lead to an accident or near miss

extended vector line: An extension of the vector with no regard to the actual direction or force of the actual vector

extremis situation: A situation, typically in regard to vessels colliding, that requires extreme maneuvering on the part of all involved to avoid disaster

eyes up / eyes down: A process that enables the captain and navigator to share who has primary lookout responsibility. Primarily used when the captain needs to break the ten-second rule to adjust or more closely examine the various information screens or other sources of information.

final diameter: The diameter of a turn measured from the vessel's position when the heading is 180° and 360° from the heading at the start of the turn

flag state regulatory body: The agency that enforces the regulations of a flag state. In the United States this function is served by the USCG.

freewheeling shaft: Allowing a jet shaft to spin freely, unclutched, as water passes through the inlet tunnel

glance down capable: A source of information—often ARPA or ECDIS—that is set up so that a maximum amount of information can be easily obtained by merely glancing at the screen. A key concept in being able to observe the ten-second rule.

global load: A load that affects the entire structure of a vessel

global structure: The framing and supports that the hull and deck are built onto

hang roll: A roll where the vessel is sluggish to return to an upright condition. Typically happens on top-heavy ships with reduced stability or righting arms.

High Speed Craft (HSC) Code: The rules governing the construction, arrangement, lifesaving, accommodation, and training required for high-speed craft operations

home position: The alignment of the jets that produces a known walking speed with no rate of turn. This baseline arrangement is used to start the process of walking.

hump speed: The speed at which planing- and semiplaning-hull vessels are supported by the water passing under the hull rather than by the principle of displacement

hydraulic ram: A piece of equipment that uses high-pressure fluid to fill a cavity, forcing the ram to either extend or retract. These pieces of equipment are capable of producing the high amounts of force required to move the buckets and backing plates against the jet wash.

ideal environment: An environment where there is no current or wind affecting the ship. While this situation is likely to be rare, it is important to consider how the ship will move without these forces when establishing the home position.

impeller: A rotating blade assembly used to push water through a tunnel

inlet duct tunnel: The passage by which water is channeled to the impeller, forming the front part of the water jet

inlet flow: The supply of water to the inlet that feeds the impeller. Insufficient flow can result in a cavitation condition.

inspection port: An opening that allows access to the front side of the impeller from inside the jet equipment room. Often the plate securing this opening is bolted to a flange to be watertight.

integrated bridge system: A grouping of the sources of information and controls needed to stand a navigation watch within easy reach and use of the watch stander

integrated docking control A control system that enables the user to control all the various controllable forces available by means of a single control, typically a joystick or combinator. When using these systems, the operator is controlling the ship's movement or resultant vector instead of the individual force vectors.

interceptor: A vertical control surface that is used to produce a high-pressure bubble under the stern to aid in getting the ship up on plane. Actuated by either a hydraulic ram or linear motor.

intersection point (IP): The point at which the extended vector lines cross. Where the resultant vector is understood to act on the vessel.

J turn: A maneuver that is most effective with a jet drive where the backing plate is moved from the full-ahead (open) position to full astern (closed) without reducing pump rpm. This causes a very high-powered astern wash to quickly stop the vessel. Smaller boats will often nearly submerge their bows. The bucket is then angled to vector the astern wash to rotate the vessel. As the vessel nears the reciprocal course, the backing plate is returned to the full-ahead (open) position, and the bucket is shifted so that the turn is completed.

jet bolt ring: The ring of bolts that attach the nozzle to the transom

jet creep: Slight forward or aft motion caused by a subtle imbalance in the forward and aft thrust developed in the zero-bucket position

jet equipment room: The room where the various equipment needed to operate the jets is located. In smaller vessels this may be the same as the engine room. Larger vessels will typically isolate the two spaces with a watertight bulkhead.

Kort nozzle: Also known as a ducted propeller, where the duct surrounds the propeller like a sleeve. Typically used to improve propeller forward-thrust efficiency.

leeway: The sideways movement of a ship

lineshaft bearing: The pieces of equipment that support and stabilize the shaft along its length between the reduction gear and the impeller

magic-box syndrome: A tongue-in-cheek description of when a watch stander spends more time looking at their information screens than out the window while trying to solve the navigation problem

maneuvering table: A document that shows the maneuvering characteristics of a vessel as determined during builder's trials. This document will list the conditions present during the tests used to construct the table. These tables are typically not reliable, since the maneuvering characteristics of water jet ships change quickly when conditions do not match those of the tests.

matched-angle walking: A method of vessel walking where the angles of the jets are mirrored between the port and starboard side. While this method is effective at maintaining control of a dedicated helmsman, it is not the most effective in terms of vessel control. In reality, the speed and direction of twisting are being controlled, not so much the walk.

maximum landing speed: The maximum speed at which the hull can be allowed to lie along the pier or fenders without risking point or global damage to the hull

maximum rate of turn: An autopilot control that limits the rotational speed a vessel is able to turn at. Often, if this maximum is exceeded, the autopilot will drop out and stop functioning. Depending on the specific system, an alarm may or may not sound when this happens.

maximum safe speed: The highest speed a vessel may transit at, considering a number of variables

maximum speed: The highest speed a vessel can proceed at

mental model: A person's expectation of how a situation will work out, on the basis of their understanding of currently available information

middle of the good water: The best place for a vessel after balancing all the various navigational considerations, such as depth, bottom profile, distance to shoal, shore facilities, channel, lane or slip width, set, and drift. Not always the middle of the channel, lane, or slip.

middle water: Water depths between shallow water and deep water. The depths at which the bow wave may reach down to the bottom, depending largely on speed. Not typically an area of concern for displacement hulls since they are not trying to plane.

minimum safe jet distance: The minimum distance that something floating on the surface can get to the ship side near the jet intake without being sucked under. This will be different for each jet arrangement and will increase with shaft rpm.

mixed-flow impeller: An impeller that produces thrust by pushing water to the outsides of the tunnel before passing into the thrust nozzle

moment knob: The integrated docking console control that is used to adjust for heading while walking or twisting the vessel into place

monohull: A vessel with one hull. The most common hull classification.

multihull: Any vessel with more than one hull. Both catamarans and trimarans fall into this category.

natural range: Any two features that line up one in front of the other that will provide a reference for movement, either fore and aft or to port or starboard

navigation picture: The situation the vessel is currently in, including surrounding traffic, position, and placement along the intended track

navigation problem: The application of the navigation picture to the current mental model to continue to move the vessel down the intended track

navigator: The person assisting the captain in solving the navigation problem and running the vessel

NFU (non-follow-up): A backup control found on traditional steering systems that bypasses the helm to directly control the steering signal going to the electrohydraulic steering system. When this system is selected, the rudder angle will not match the position of the NFU control.

nondisplacement pump: A pump that is not capable of pulling liquid in so that it can be pumped. These pumps require a certain amount of feed pressure to prevent becoming air-bound. Typically used when high volumes of low-viscosity fluid are being pumped.

no-operation zone: The area on a cavitation table above the pre-solved cavitation limit for various speeds through the water and shaft rpm combinations. Continual operations in this zone will quickly cause significant damage to the impeller.

north up: ARPA and ECDIS display setting that keeps north to the top of the screen at all times, regardless of vessel heading

nozzle: Part of the water jet after the impeller that focuses the jet wash to increase the overall power of the jet wash

off-course alarm: An autopilot alarm that indicates when the system can no longer keep the vessel on course within the prescribed limit

on step: On plane, a hull that is partially or fully supported by the water passing underneath rather than by the displacement of the surrounding water

OODA loop (observe, orient, decide, act, then repeat): Technique that searches for new information that may change the mental model and, if not, causes action to correct

ordered speed: The speed through the water the current shaft rpm should produce

ownship centered: An ARPA and ECDIS setting that keeps the origin point of an EBL/VRM at the ship instead of offset

pier heading: The true heading of the pier or dock during a maneuvering situation. This may or may not be the same as the slip heading. Both are valuable when maneuvering the ship, to reduce errors of point of view. Keeping the vessel on pier/slip heading typically makes a maneuvering situation easier.

pinched bucket: Used to define when a bucket is not in one of the three main positions: full open (ahead), full astern (closed), and zero (balancing the force of ahead versus astern). Typically used with ahead or astern, "pinched bucket ahead" indicates that effective wash may not be being achieved.

pinning to pier: The process of directing the resultant vector toward the pier, against the fenders, to hold the ship in position while lines are being worked. Also used to finely tune the shaft rpm prior to slacking lines while getting underway.

pivot point (PP): The geographic center of the water plane of the hull. The point where the ship pivots around. This point will move forward or aft along the keel and is not the same as the point of rotation.

planing: The condition where either a planing or semiplaning hull is supported by the water passing under the hull, rather than by displacing the surrounding water

planing hull: A hull that is designed to be fully supported above the water to reduce wetted surface as much as possible. Often their waterlines can be seen to physically rise up out of the water once planing and will drop back into the water once below hump speed.

point load: A load or force that affects only a small portion of the hull, such as a protrusion from the pier that is contacted before a fender. Typically, forces such as these will puncture the hull or bend nearby frames, with no other damage to the ship.

point of rotation: The point at which the ship rotates about; best seen as the point where there is no motion, only rate of turn. This point can be anywhere on or off the vessel. Typically, the farther this point is from the pivot point, the higher the vessel's speed is, excepting the presence of an outside force such as set and drift, or a tug.

power bias: The direction of overall thrust in split plant. An ahead bias means that with the power plant split there is more overall ahead power; given time, this will result in headway.

power pack: The sump, filter, motor, and pump assembly for a hydraulic system

predicted true vector: The intersection of the ownship-centered EBL and VRM, set to either the current or proposed six-minute distance. Primarily used to solve the navigation problem in real time rather than using the trial maneuver function on ARPA. Also useful to determine the course to steer while proceeding along the track.

prime mover: The source of rotation for the drivetrain. Can be a motor, engine, or turbine; typically for water jet ships, high-speed diesels or gas turbines are used.

proactive decisions: Choices that are made from a well-formed mental model in a timely fashion

PTO (power takeoff): A shaft that is rotated by another shaft for a different purpose. In this case it is meant to describe a shaft-driven hydraulic pump that is rotated by the impeller shaft.

pump housing: The part of the tunnel that surrounds the impeller. Tolerances between the housing and impeller blade tips are typically measured in a 1,000th of an inch.

pump overspeed: A condition mainly associated with nondisplacement pumps where the speed of the pump exceeds the fluid flow rate to the pump. Pumps in this condition are referred to as air-bound, or in a cavitation state. In either case, a pump in overspeed will not produce effective thrust, suffering thrust breakdown.

pumping a jet: The process of sharply increasing power to a bucket at hard over to increase rate of turn. Properly done, power is reduced as soon as any change in speed is seen, thereby keeping the point of rotation the same but increasing ROT. Care must be taken, since water jet ships will typically accelerate more quickly than traditional ships.

pumping a rudder: The process of sharply increasing power to a propeller ahead of a hard-over rudder to increase rate of turn. Power is reduced in such a way as to minimize changes in speed but keep ROT high.

questioning attitude: An essential part of chronic unease or OODA looping where the watch stander is always looking for new information to be used in revising the mental model

radar overlay: An information layer of ECDIS that displays the raw radar image on top of the navigation chart. This allows for instant correlation of GPS- and radar-derived positions. Any offset of the radar image from the chart indicates a loss of position accuracy in one of the two systems.

reactive decision: A choice made in an instant with little to no mental model or situational awareness, often made as a last-chance effort to avoid an extremis situation. Can be a link in the error chain and an indicator of proceeding above maximum safe speed.

reduction gear: The drivetrain gears used to reduce the shaft rpm from the prime mover to a useable rpm for the impeller

relative-bearing drift: Movement to the right or left of an object in the bridge window, best observed from a stationary position

relative vector: An ARPA presentation of another vessel's motion that will match how they move in the window. The reference point for all vectors in this mode is ownship; therefore ownship will not have a vector.

resultant vector: The product of combining the direction and power of two or more vectors, in this case the ahead and astern jet vectors. The resultant vector can be found graphically by placing the tail of the second vector at the tip of the first. The resultant is the third side of the triangle.

reverse deflector: An element of the bucket that directs the wash from the backing plate under the vessel. Can be combined with the backing plate.

ride control: A system that makes use of active underwater surfaces to reduce the effects of the sea. This system is typically automated and can make use of trim tabs, interceptors, foils, and wings. Proper operation of this system makes getting up on plane easier and helps ease the global loads on the ship.

rooster tail: The plume of water that leaps up behind the ship when up on plane. Water jet vessels often have large rooster tails, since the jet wash adds to the effect. For large ships these tails can be as tall as 20 feet or more.

round turn: A man-overboard-recovery turn where the ship is held in a continuous, high-steering-angle turn for a full 360 degrees but does not take into account cavitation management. The victim will normally be on the bow outside the turn when back on original course.

safe-haven port: A port of refuge that can be used by an HSC vessel in the case of an emergency or diversion. Most often used for a medical emergency, or when the safe-operating envelope has been exceeded.

safe-operating envelope: The sea conditions that when exceeded lead to an increasing level of danger to an HSC ship and her crew. These conditions will frequently be less severe than what traditional displacement ships can handle.

safe working load: The stress a piece of equipment can expect to regularly tolerate without failing, often five times less than the breaking strain

semiplaning hull: A hull that while supported by the water passing underneath does not visually rise up on top of the water while on plane

shaft rpm counter: The indicator or rotations per minute of the shaft

shaft seal: The piece of equipment around the shaft that prevents water from entering the jet equipment or engine room

shallow water: Depths of water where the closeness of the bottom to the hull will induce shallow-water effects, such as squat. Planing is not typically possible or advised in these depths.

shallow-water effects: The various ways that increased speed in shallow water affect the vessel, such as squat, increased directional inertia, increased speed inertia, bow cushion, and stern suction

single-jet walking: A method of vessel walking where the angle of one jet is used to control heading while the other jet remains static. While this method is effective at controlling the vessel as a whole, it is not as effective at maintaining control by a dedicated helmsman. Vessel speed is controlled by the same jet, simplifying the process of walking for the operator.

situational awareness: Using all available information to form a mental model that can be used to solve the navigational problem. Knowing what is going on around the ship.

six-minute distance: The distance the ship will go in six minutes or one-tenth the current speed

skills atrophy: The ability level in a given skill will fall off when not regularly practiced.

snap roll: A roll where the vessel quickly returns to an upright condition. Typically happens to ships that have low centers of gravity or large righting arms, like those found on multihulls.

soliton wave: A power wash of water that is produced by all ships moving at or near hull speed. Solitons are a particular problem for high-speed vessels, since they have sufficient power to create large, damaging solitons in places where traditional ships might not.

speed log: A sensor that shows the speed over ground or through the water of a vessel, depending on how the measurement is taken

split plant: A special drivetrain arrangement for water jet ships where half—one side—of the buckets are set to produce ahead wash, while the other side is set for astern wash. Higher rpm on the astern jets will allow increased ahead rpm, producing effective wash for steering with an ahead bias. Using this plant arrangement prevents water jet ships from bare-steerage problems when maneuvering. Most effective for vessel speeds of 5 knots or less.

squat: The tendency of the stern to settle in shallow water when speeds are increased, due to the Bernoulli principle. This tendency is magnified by the low-pressure areas at the stern caused by jet inlet suction. In certain cases it is possible for a squatting water jet vessel to suck the stern down to the bottom.

static jet: The jet that remains largely in a fixed position when maneuvering under the single-jet method

stator bowl: Part of the nozzle that holds the inner and outer nozzles apart

stator vanes: The individual elements that maintain the spacing between the inner and outer nozzles. Also used to remove the twist imparted to the jet wash by the impeller.

steering-control system: A computer that takes the steering commands from the helm and processes them to send steering signals to the electrohydraulic steering in the jet rooms

steering lever arm: A lever arm that exists between a rudder or bucket and the keel. Also used to describe the lever that theoretically exists between the IP and the PP when a resultant vector is acting on the PP.

steering reverse: A control mode made possible by the steering-control system that inverts the steering signal going to one side of the steering buckets. This enables the buckets to mirror each other across the keel instead of matching each other. Use of this mode allows the operator to easily twist the ship without making use of the docking station. This mode will also allow walking, but at a much-reduced level of control compared to docking-station or combinator controls.

stern wave: The wave of water that comes off the stern. This wave is often a standing wave aft of the ship and will make up part of the rooster tail. This wave can also be referred to as the transverse wave, which, when crossing the bow wave at nearly right angles, has the potential to produce damaging waves.

stiff ship: A stability condition where the right arm is large, leading the ship to quickly return to an upright orientation

stop-and-twist turn: A man-overboard-recovery turn that uses vectored water jet thrust to stop and then twist in place before proceeding back along a reciprocal course to rescue the victim. This turn reduces the overall distance from the victim while providing a quick maneuver that takes into account the additional requirements of cavitation management that are not present on traditional ships.

success chain: A series of actions and choices that when consistently taken seek to reduce the odds of an accident happening. The opposite of an error chain.

synchronous roll: A roll where the ship's deck is always parallel to the surface of the ocean. This is a dangerous situation that if left unchecked could result in a capsize.

tender ship: A stability condition where the righting arm is reduced, leading to the ship slowly returning to an upright orientation

ten-second rule: The maximum amount of time the captain can look down from the window and still be said to be keeping a proper lookout

Texas chicken: A carefully coordinated maneuver that makes use of bow cushion and takes into account stern suction, to keep two meeting ships apart when passing in a narrow channel

thinking with the throttle: The act of reducing power simply because new information was presented, without taking the time to evaluate the importance or immediacy of that information. Typically seen as an indicator of proceeding above maximum safe speed.

thrust bearing: The lineshaft bearing that imparts the thrust on the shaft into the hull of the ship, producing forward or aft speed for traditional ships. Water jet ships use this bearing to keep the impeller properly positioned in the pump housing.

thrust breakdown: A condition of reduced thrust commonly caused by cavitation or pump overspeed. Correcting this condition is usually a matter of reducing rpm and repriming the pump.

thrust nozzle: Part of the water jet after the impeller that focuses the jet wash to increase the overall power of the jet wash

thrust vectoring: Directing the thrust of a water jet at an angle from the keel to produce other than strictly forward and aft motion

toed in: Angling a jet so that it is pointed toward the keel

toed-in spin: A spin where the plant is split in even power and all the jets are pointed toward the keel

toed out: Angling a jet so that it is pointed away from the keel

toed-out spin: A spin where the plant is split in even power and all the jets are pointed away from the keel

track mode: A form of autopilot that controls not only heading but also vessel speed, in such a way as to follow the track laid down by the bridge team. Extreme caution should be exercised when using this mode.

transition: A significant change in the motion of the ship in a maneuvering situation

transverse wake: The horizontal stern wave across the stern of the ship that travels at or near the same speed as the vessel

trial maneuver: An ARPA function that allows the operator to experiment with ownship course and speed changes to see the change in the navigation picture prior to taking action. Use of this mode requires setup time and has the potential of forgetting that it is engaged, possibly leading to an ARPA-induced collision.

trim tabs: Plates placed at the corners of the transom that push down on water exiting from under the hull to lift the stern and provide stability and a more level trim. When active they can be part of a ride control system.

trimaran: A vessel with a large, narrow center hull flanked on both sides by shorter, shallower narrow hulls. These outer hulls are connected by a bridging structure, although this bridge is not as flat as those found between catamaran hulls.

true presentation: An ECDIS display setting that allows all vessels underway, including ownship, to move across the chart, as would be seen by a bird hovering over the ship in real life

true trail: An ARPA feature that displays the true, last position of a contact on the plotter in the form of an extended echo. A very good tool for seeing a target's real course when using relative vectors.

true vector: An ARPA display presentation of another vessel's motion that will match how they are moving across the ocean when viewed from above. Ownship will also have a vector that is used against the predicted true vector for solving the navigation problem.

turning diameter: The diameter of a turn for a given steering angle as measured from the ship's center of gravity

turning lever arm: The lever arm developed between the pivot point and the steering or turning force

twisting: Causing rate of turn by setting the direction of the propellers or buckets against each other. Due to the short steering lever arms, this tends to be a slow maneuver on traditional ships. Water jets can increase ROT by toeing in or out.

variable range marker (VRM): A function of radar that allows the user to lay a ring around the ownship or user-defined offset origin

visual-bearing repeater: A compass card mounted so that visual bearing can be taken by the bridge team for position fixing. Also useful for ascertaining the bearing drift of an object.

VRM safety bubble: Typically a 0.25 NM ring centered on ownship that is used with relative vectors to prevent extremis situations from happening

walking: The process of using the resultant vector on the pivot point to move a vessel laterally while maintaining pier heading

walking window: The angular range of the jets that allows heading control while walking without toe-in or toe-out twisting

watch engineer: The engineering person in charge of the propulsion plant

water pickup tubes: Tubes that extend into the jet wash so that water can be forced back without the need of an additional pump

wave train: The organized sequence of waves in the open ocean

wet deck: The underside of the bridging structure that connects multihulls together

wet-deck slam: A condition where the crest of a wave or swell slams into the wet deck, causing extreme global loading and vibration

Williamson turn: A turn that uses specific rudder angles and wheel-over points to bring a ship back on her reciprocal course for recovery of a man overboard, but does not take into account cavitation management

zero bucket: The bucket position that splits the jet wash to balance ahead and astern forces, not flow. Similar in nature to the zero pitch of CCPs.

zero speed through the water (ZSTW): A condition where the ship has no speed in any direction through the water

BIBLIOGRAPHY

The following are additional references or sources of information for this book.

Barber, Capt. James A. *Naval Shiphandler's Guide*. Annapolis, MD: Naval Institute Press, 2005.

"Bill Hamilton Founder" (2019). Retrieved from http://hamiltonjet.com/global/sir-william-hamilton.

"Boat with Propelling Turbine." *Scientific American Supplement* 38, no. 980 (October 13, 1894): 15659.

Cauvier, Capt. Hugues. "The Pivot Point" *The Pilot*. no 295 (October 2008)

Gagliano, Joseph A. *Shiphandling Fundamentals for the Littoral Combat Ship and the New Frigates*. Annapolis, MD: Naval Institute Press, 2015.

Hagopian, Jeff. Conversations with Hagopian, a US master mariner, water jet captain, and water jet shiphandling instructor.

Hollesen, Henrick. Conversations with Hollesen, a Danish master mariner, water jet captain, and water jet shiphandling instructor.

INCAT type-rating program for the INCAT 98 m vessel.

International Navigation Association. *Guidelines for Managing Wake Wash from High-Speed Vessels*. Report of Working Group 41 of the International Maritime Navigation Commission. Brussels: PIANC General Secretariat, 2003.

Lyons, Bill. Conversations with Lyons, a US master mariner, water jet captain, and water jet shiphandling instructor.

MacElrevey, Daniel H. *Shiphandling for the Mariner*. 5th ed. Atglen, PA: Cornell Maritime Press, 2018.

Maritime & Coastguard Agency, prod. *High Speed Craft—Control in Following Seas*. Southampton, UK: BMT Seatech, 2005.

Owen, Paul. *High Speed Craft, A Practical Guide for Deck Officers*. London, England: The Nautical Institute, 1995.

Weeks, Bud. Conversations with Weeks, a captain in the US Navy (ret.), and director of NSS, US Navy.

INDEX

AUTHOR'S BIO

John Kinkela was raised in New Jersey and attended SUNY Maritime, graduating with a BS in marine environmental science and a 3rd Mates Unlimited license in 2001. Between 2001 and 2008, John sailed on various ship types, traveling all over the world to obtain an Unlimited Master Mariner's license in 2008. Also in 2008 John completed the High Speed Vessel (HSV) 2 *Swift* type-rating process with the INCAT training team and joined the crew of HSV 2 *Swift* as chief mate. As a member of the crew, John made numerous high-speed Atlantic Ocean crossings and multiple transits through both the Suez and Panama Canals. On these voyages, numerous port calls were made along the western coast of Central and South America, along a majority of the African coast, and at ports in southern Europe. During these port calls, John was a key member of the maneuvering team, either serving as primary lookout or directly handling the ship from the docking station. After sailing for 14 years, eight of those as chief mate, John came ashore to teach high-speed navigation and water jet shiphandling for the US Navy's Littoral Combat Ship. John has been an instructor at the US Navy's Surface Warfare Officer's School in Newport, Rhode Island, since 2014. John lives with his wife and three children on the Jersey Shore near Asbury Park and can be contacted by email at waterjetguide.jjk@gmail.com.